Contents

	Acknowledgements	vi
	Foreword	vii
1	The warmer, wetter world	1
2	How the rising seas hurt	36
3	Third World damage	58
4	Rising seas and the rich	71
5	Small states and rising seas	92
6	Facing the rising seas	107
7	Living with the rising seas	126
	Bibliography	147
	Index	150

Case studies

The Maldives	32
Australia's Barrier Reef	56
Bangladesh	68
USA	87
The Netherlands	89
The Pacific	105
Guyana	124
Canada	145

Acknowledgements

The idea for this book came from the Commonwealth Secretariat. The concerns of Commonwealth Governments about climate change and sea-level rise led to the Report "Climate Change: Meeting the Challenge" by a Commonwealth Conference on Sea-Level Rise convened on the initiative of the Government of the Maldives in Male in November 1989. I am grateful to the Commonwealth Secretariat for the cooperation it has given me and Earthscan in the preparation of this book and particularly to Dr B. Persaud, Dr V. Cable and Mr C. Krishnan of the Economic Affairs Division for their collaboration in making available the valuable papers and studies the Secretariat had prepared or commissioned on the subject of sea-level rise and in providing advice on this exercise. The Secretariat remained enthusiastic even when I suggested a book which would cover both member and non-member countries of the Commonwealth. I am grateful to Neil Middleton of Earthscan for suggesting me as the book's author. Practical assistance has come from sources too numerous to name, but I am especially grateful to Phil O'Keefe of Newcastle-upon-Tyne Polytechnic. The book comes to Vicky, with love.

THE RISING SEAS

Martin Ince

Earthscan Publications Ltd   London
With the co-operation of the
Commonwealth Secretariat, London

First published by
Earthscan Publications Ltd
3 Endsleigh Street, London WC1H 0DD

British Library Cataloguing in Publication Data
Ince, Martin
  Sea-level rise.
  1. Coastal regions. Environment. Effects of rise
  in sea level.
  I. Title
  333.784
  ISBN 1–85383–077–1

Production by David Williams Associates 081–521 4130
Typeset by Selectmove, London
Printed and bound by Cox & Wyman Ltd, Reading

Earthscan Publications Ltd is a wholly owned and editorially
independent subsidiary of the International Institute for
Environment and Development.

Cover photographs by Richard Smith/*The Independent*

Foreword

by the Commonwealth Secretary-General

There are few environmental issues which have captured popular fear and imagination more than the threat of climatic change and rising sea levels. The prospect of whole countries disappearing beneath the seas is a vividly dramatic illustration of the chilling consequences of human activity which has neglected the environment.

Many Commonwealth countries are threatened by climate change and its consequences. Some large countries are involved but the fact that nearly half of Commonwealth member countries are small island states, which are greatly exposed to rising seas and storms, makes this issue one of special significance for the Commonwealth.

At the Commonwealth Heads of Government meeting in Vancouver in October 1987, President Gayoom of the Maldives described the way in which unprecedented waves had caused great destruction in his country. He posed the question "Is this the beginning of sea-level rise consequent upon global warming?". His enquiry subsequently led to the establishment of a Commonwealth Expert Group under the chairmanship of Dr Martin Holdgate, Director-General of the International Union for the Conservation of Nature and Natural Resources. The Report of that Group was presented to Heads of Government in Malaysia in October 1989 and was an influential input to their discussions. They adopted the Langkawi Declaration on Environment – a programme of action to address several environmental threats, including climate change and sea-level rise.

The threat of global warming and consequent sea-level rise has progressed from an alarmist scientific speculation to more securely based evidence of current change and future trends. Thanks to the work of the UN Intergovernmental Panel on Climate Change, there is now a reasonable scientific consensus on the phenomenon of sea-level rise and its future magnitude.

Although much more research is needed, analysts have been able to look in some detail at what this could mean for particular low-lying countries and communities. As part of the work of the Commonwealth Expert Group, a series of valuable case studies were commissioned, financed in part by the Australian Government. I am glad that this book brings some of this material to a wider audience.

How should the world prepare for sea-level rise? Sea defences are one option but applicable only to those countries that can afford the considerable resources involved. Some form of adaptation of life-styles might be possible in the threatened Pacific and Indian Ocean islands. But realistically, substantial and continuing sea-level rise would threaten the viability of whole nation states. The prospect is a totally unacceptable one. It is for this reason that the international community must find ways to prevent further growth of the greenhouse gas emissions which cause the problem. Commonwealth leaders have called for the early conclusion of an international convention to protect and conserve the global climate under the UN's auspices. The World Climate Conference in Geneva later this year will be a major opportunity for the world to develop a co-ordinated response to global warming and its consequences. It is to be hoped that international negotiations will begin soon and culminate in an agreement which is effective and equitable. Rich countries, who account for the major part of greenhouse gas emissions, are beginning to realize that they have an obligation not only to their own citizens but also to the world at large to exercise much greater restraint over energy use and noxious chemicals. But poor countries must also co-operate to find solutions.

The challenge to policy makers is an enormous one. But they must keep in mind the consequences of inaction. This book endeavours to describe what those consequences might be, particularly for some of the world's more vulnerable communities, and how they might be addressed.

Emeka Anyaoku
August 1990

1 The warmer, wetter world

Storms send the sea into houses and across fields in Britain; high tides make hundreds of thousands homeless in Bangladesh; low-lying islands in the Pacific no longer produce food as rising salt water attacks their crops; the huge delta of the Mississippi spreads into areas that used to be dry land. All over the world, coasts seem to be slipping beneath the waves.

This book will look at the silent disaster of the rising sea, which threatens thousands of square kilometres of land around the world, the livelihoods and ways of life of millions of people, and the future of unique species and biological habitats. It will look especially closely at the likelihood that the seas are rising because the Earth is getting warmer, a process which may be due to the effects of human activity on its atmosphere. But the seas have always risen and fallen, as a pleasurable day out at the coast will show.

Take a walk along the shore in the Highlands of Scotland, or in many other parts of the world, and you may find that you can stroll on the beach with no risk of getting wet. Under your feet, as you walk along tens of metres above sea level, you will find sand, complete with seashells. This is a raised beach, a relic of times, thousands of years ago, when the sea was far higher than it is today. These raised beaches, in common with other evidence from many parts of the world, speak eloquently of the fact that the relationship between land and sea has been constant only in its variability throughout the history of the Earth. Many coastlines exhibit raised beaches, while in other parts of the world underwater investigations have revealed shorelines, river channels and other land features far below today's low-water mark. In some cases, there is strong evidence that major changes in sea level have occurred over very short periods of time.

But the fear today is that the activities of the human race are leading to changes in the Earth which will mean sea level changes far faster than those which nature manages without artificial assistance. This book aims to look at the possibility of such changes, consider how they might affect different parts of the world and the people who live there, and assess the new policies and new knowledge which will be necessary to deal with the problem.

I shall argue throughout that the real issue about the rising sea is not how high up the beach the high tide comes. Instead, the subject contains subtle and difficult arguments about science, money, morality, people and politics. In any case, the sea level issue is not a free-standing policy problem. Instead it is a part of a more general issue: the extent to which gases emitted to the atmosphere by industry and other human activities will raise the Earth's temperature by trapping incoming energy from the Sun which would otherwise be radiated back into space.

This is the greenhouse effect, and the gases that cause it are the greenhouse gases. As we shall see, countless uncertainties are associated with the greenhouse effect. To these is added a further layer of questions about just how the greenhouse effect might, as it unfolds, be translated into rising sea levels. The link between the two involves a complex chain of cause and effect involving oceanography, glaciology, meteorology and other messy subjects which we shall encounter shortly.

But of more significance than these scientific uncertainties, doubled and redoubled by the unknowns in our current knowledge of the way the Earth works, is a series of even more intractable issues about the way the world works. Scientific knowledge leaves large areas of doubt about how much sea levels are going to rise, but a rise would not occur in a uniform manner throughout the world. Different types of coastline would react differently to an equivalent rise in the sea, which might leave a rocky shore unaffected and cause severe damage to a sandy one, or flood a populous estuary area while leaving an area of cliffs all but unaffected.

More importantly, the rising sea will affect different areas of the world differently in terms of its impact on the people who live there. It will do so simply because people themselves vary, and most particularly they vary economically and in how they live. For example, London is protected from flooding by the Thames Barrier, installed because of a growing flood hazard caused mainly by the gradual subsidence of South-East England. Many Third World areas face far more immediate and dangerous flood hazards, but they do so without a Thames Barrier because they have neither the wealth nor the political influence to buy one. If sea levels rise, the Barrier will be of even more use to London, and places without one will be even more disadvantaged economically.

It follows that the key factor determining how the rising sea will affect any given individual is that individual's existing social and economic circumstances. If you are poor already, rising sea levels will make you poorer by removing your livelihood and the place where you live. If you are living somewhere that is prone to pollution, flooding or other forms of disaster, rising sea levels will tend to make your problems worse. For example, the rising sea will flood your land more often, poison your land and drinking water with salt, and damage the drainage and flood defence system which defend you from the sea.

In richer countries the same problems will arise in some of the same ways, but their effects on the societies of the developed world will be quite different. In the extreme case of the complete destruction of livelihoods by the rising sea, the nationally and personally wealthy states and individuals of the developed world would be far better prepared to cope by re-creating the lives of the affected individuals elsewhere. In the intermediate case – damage short of complete destruction – developed world countries would be able to spend far more on sea defences, new drainage and sewage systems and other measures used to counter rising seas.

In addition, it so happens that the countries of the world whose existence is most fundamentally threatened by the rising sea are all in the Third World, with the exception of the Netherlands, a country whose existence is entirely dependent upon sea defences. Across the Pacific and into the Indian Ocean lie a clutch of nations like the Maldives (one of the Commonwealth nations whose concern with the rising sea has prompted a variety of initiatives on the subject, including this book), whose land is all close to sea level and which risks flooding, total immersion or other forms of damage as the sea rises.

This book's analysis will maintain that in an important way, rising sea levels do not alter the basic facts about the world. However, the rising sea is not a zero-sum game like commodity speculation, in which some participants get richer at the expense of others. Like many other environmental problems, the rising sea will impoverish virtually everyone it touches, but will do so without creating a separate social or economic group who win out as the other loses. Large areas of developed world countries like the USA and Britain are threatened by the rising sea. On the southern and eastern seaboards of the USA, rising sea levels – not associated with the greenhouse effect – are already causing widespread destruction. Sea level rise has the power to make rich areas poor, put physical stresses on areas accustomed to immense physical advantages, and create immense demands for sea defences and other investments which are likely to be costly even for developed nations.

The problem for anyone interested in sea level rise is to understand just how it might occur, on what scale and how fast. The third part of this question is especially important to the inhabitants of land areas threatened by sea level rise, since a change which would be disastrous if it happened overnight or over a year might be bearable if some years were available to find out about the problem, think about its solution, implement it – and, of course, pay for it.

The starting point for understanding these problems is the greenhouse effect itself. Since the early 1980s, when the Office of Technology Assessment of the US Congress published a report pointing out that the sheer amounts of carbon dioxide being pumped into the Earth's atmosphere were bound to have some effect on the climate, shelves and, indeed, whole libraries have been filled with material on the greenhouse effect. The subject has been widened out with the discovery of new ways in which the effect might take place and by the application of sophisticated scientific modelling processes to the problem, along with a whole gamut of work on the Sun's radiation and the way in which it interacts with the Earth.

But first things first. What is all this talk of greenhouses? And why do they have an effect named after them?

Greenhouses allow plants which would have trouble growing out of doors to be grown under cover instead. To some extent they do this by cutting out the winds and frosts which would affect plants out of doors, but the greenhouse effect proper comes into play when the greenhouse receives radiation – mainly visible sunlight – from the Sun. Most of the radiation emitted by the Sun is light, at wavelengths visible to our eyes – or, to put it the other way round, we have evolved to see things in the type of light which is most available. When this radiation reaches the greenhouse, it goes through the glass and warms the inside.

The radiation emitted by any object is characteristic of its temperature – any physicist looking at the amount of energy emitted by the Sun at different wavelengths could tell you that its surface is at a temperature of about 5600°C. Naturally even the most efficient greenhouse is not this warm, so the energy which the things in it emit has a quite different set of wavelengths. The tomato plants, the greenhouse floor and everything else inside it emit energy of longer wavelengths, invisible to human observers. And just as visible light will not travel through brick, the radiation given off by objects at room temperature will not travel through glass, the main substance of which greenhouses are made. The heat builds up inside the greenhouse instead, keeping everything in it warm. Hey presto – the greenhouse effect!

In the case of the Earth and its greenhouse effect, no sheets of glass are necessary. Instead, the contents of the atmosphere act to hold in energy which would otherwise be lost to outer space after being reflected from the Earth's surface. The gases in question do not do this by forming cloudy layers which reflect heat back to the Earth – a perfectly clear sky can be as efficient a greenhouse layer as an overcast one. This is because all the greenhouse action takes place at wavelengths which our eyes cannot see, in the infra-red, energy transmitted at slightly longer wavelengths than visible red light. The chemical bonds which hold atoms together into molecules often involve energy in these wavelengths, which means that the many molecules can absorb infra-red energy just as glass does.

The greenhouse effect is not a completely artificial one. The main greenhouse gas in the atmosphere is carbon dioxide, which is present naturally – although human activities are now adding hugely to the amounts of it and the size of the greenhouse effect it causes. The greenhouse effect is seen in its fullest form on the planet Venus, whose surface temperature is some 400°C higher than it would be if the planet had no atmosphere. The dense Venusian atmosphere consists of 95 per cent carbon dioxide, although other gases present in small amounts are also responsible for a significant part of the greenhouse effect there.

The same applies on Earth, where gases which are present in far smaller amounts than carbon dioxide, but have a stronger greenhouse effect per molecule, are also important. They include methane (the "natural gas" used by domestic consumers and businesses), ozone (a form of oxygen where every molecule has three atoms instead of the usual two), chlorofluorocarbons (complex organic compounds involving carbon and the halogen elements chlorine and fluorine), and several oxides of nitrogen. Many of these gases are a cause of environmental concern even if their role in the greenhouse effect is disregarded. Ozone is a poison, nitrogen oxides are thought to be carcinogenic, and the chlorofluorocarbons are responsible for the destruction of the ozone in the upper atmosphere which prevents harmful ultraviolet light from the Sun reaching the Earth.

However, working out exactly how large a greenhouse effect these gases might cause is a complex task involving a large number of uncertainties. First there is the most significant of the greenhouse gases, carbon dioxide. Carbon dioxide is not a massively powerful greenhouse gas, molecule for molecule; as we shall see, plenty of less common gases are proportionally far more significant in the greenhouse effect. But carbon dioxide dominates the greenhouse effect because it is much the most common of the greenhouse gases. More problematic still is the fact that this odourless, colourless and completely non-toxic gas is the characteristic emission of twentieth-century industrial society.

Carbon dioxide is emitted every time a piece of wood is burnt in a stove in the Third World – or, for that matter, every time an animal breathes out. Comparatively small quantities such as these can easily be absorbed by plant life – everything from trees to plankton in the ocean – which takes in carbon dioxide and emits oxygen. The Earth's atmosphere as we know it – consisting mostly of nitrogen plus just over 20 per cent oxygen – was initially produced, several hundred million years ago, by just this mechanism: when plants appeared which could turn carbon dioxide, which had previously been present in the atmosphere in large amounts, into oxygen as a by-product of the process known as photosynthesis, in which solar energy and a catalyst called chlorophyll are used to build plant material.

However, the real problem is that natural emissions of carbon dioxide have been added to in recent decades by large amounts of unnatural carbon dioxide produced by burning fossil fuels – oil, coal and natural gas. Although some other forms of energy, like nuclear power, are in use in modern industrial societies, most of the developed world today runs on fossil fuels. There are disagreements about just how large the world's fossil fuel reserves are and how long they might last, but the basics are not in doubt. On present form, fossil fuels which have been built up over hundreds of millions of years are set to be consumed over a period of a few centuries, a disparity with which the carbon-dioxide-absorbing machinery of the Earth will probably be unable to cope.

In addition, there is another characteristic activity of the late twentieth century which adds greenhouse carbon dioxide to the Earth's atmosphere – the cutting down of forests, especially large areas of tropical forest, for use as agricultural land. Even if farmland replaces the forest completely – which can often fail to happen – the plants on that farmland have nothing like the forest's carbon-dioxide-absorbing capacity. If the trees are burnt to clear the land, turning their carbon into carbon dioxide, so much the worse.

The sheer amounts of carbon dioxide emitted to the atmosphere by human activities are startling. Generally accepted estimates presented in London at a 1989 conference on climate change held for the benefit of Prime Minister Margaret Thatcher and other senior ministers in the UK government put annual emissions at 5.6–5.7 billion (thousand million) tonnes for 1988. For the previous decade the average annual total was some 5.3 billion tonnes. This is the total due to fossil fuel burning and cement manufacture, in which limestone is incinerated, with carbon dioxide being emitted as a by-product. In addition, more carbon dioxide is emitted by non-commercial fuel burning, mostly the burning of fuelwood in the Third World, which is counted only patchily by the statistics. Land use changes – mostly the removal of forests – account for another 1.5 billion tonnes a year on average for the decade to 1988. This figure is open to doubt and probably varies widely from year to year, but – at over a quarter of the figure for the total emissions from fossil fuel burning – is still extraordinary.

This amount of carbon dioxide adds up to about 7 billion tonnes a year. If it were all collected together at sea level and a temperature of zero centigrade, it would add up to 5600 cubic kilometres of gas, a block 18 kilometres long on one side. Even in a volume as massive as that of the Earth's atmosphere itself, this amount of pollution makes itself felt.

The amount of fossil fuel burnt has been steadily increasing since the late nineteenth century, and fortunately for all concerned, there are also accurate records of the effect of the resulting carbon dioxide emission on the Earth's atmosphere. One consistent set of records on the matter comes from an observatory on Mauna Loa, a volcano on Hawaii in the mid-Pacific. Here data on carbon dioxide concentrations in the atmosphere have been collected on a consistent basis since 1958, in a site so remote from industrial activity that we have a reasonable guarantee that the information obtained reflects world trends rather than local factors.

These data show that between 1958 and 1985 the share of

Figure 1.1 Measurements of carbon dioxide concentrations in the air from Mauna Loa, Hawaii

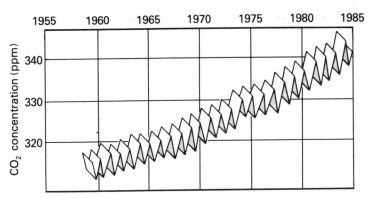

Source: UNEP

carbon dioxide in the Earth's atmosphere has increased from about 0.031 per cent to 0.034 per cent. This does not sound much, but the data are very accurate and show a consistent upward trend. There is even a visible up-and-down movement to the curve between December and June, as fossil fuel use rises in winter in the Northern hemisphere, home of most of the developed world, and falls again in summer.

To add force to the belief that industrial society is responsible for the emissions, it is possible to look at an even more arcane source of information – one which may itself be removed from the record by global warming. This is the record of the carbon dioxide trapped in air bubbles in ancient snow on glaciers and icecaps. Drilling into these old snowfalls in a highly controlled manner allows the ice in them to be dated precisely. Measuring its carbon dioxide content shows that in 1750, before the mass use of fossil fuels began, the Earth's atmosphere contained about 0.028 per cent carbon dioxide, an amount usually expressed as 280 parts per million. In other words, the whole of the industrial activity of modern times has added about a quarter to the concentration of carbon dioxide in the Earth's atmosphere.

Since the start of the Industrial Revolution, carbon dioxide emissions have been on the up and up. Figures produced by the UN Environment Programme (UNEP) point out that between 1860 and 1910, the rate of growth was consistent at over 4 per cent a year. Disturbances ranging from oil price rises to world wars and recessions disrupted the steady growth pattern after 1910 – and if fossil fuel use had gone on growing at the previous rate, carbon dioxide from that source alone would now be emitted at a rate of 16 rather than 6 billion tonnes a year.

However, it is a mistake to imagine the atmosphere as a passive bucket into which carbon dioxide is poured for long-term storage, there to remain inert except for causing an increase in the green house effect. The atmosphere itself is an active system which also interacts with the Earth's surface and its oceans. For example, the fact that the oil, coal and gas deposits exist in the first place shows that massive amounts of carbon are trapped inside the Earth. Such resources are called fossil fuels not because they are old but because they are the remains of decomposed plant life, and show graphically that carbon which was once in the Earth's atmosphere can be trapped for long periods in the Earth itself.

The same applies on an even larger scale to the Earth's massive deposits of limestone, a rock consisting mostly of carbon compounds of the metals calcium and magnesium. Trillions of tonnes of carbon dioxide are locked away in limestone which would otherwise cause a severe greenhouse effect: indeed, the main difference between the Earth and Venus is that there have never been Venusian seas in which limestone could form to suck carbon dioxide from the atmosphere.

There are many estimates of the future for the atmosphere's carbon dioxide concentration. As we shall see later in this book when we look at policies for dealing with global warming, such estimates can be produced on a wide range of assumptions, from massive growth in the world economy propelled by vast increases in fossil fuel use, to recessions or to "green growth" futures in which energy efficiency and renewable energy dominate. But if

Figure 1.2 **Global warming in 2030**

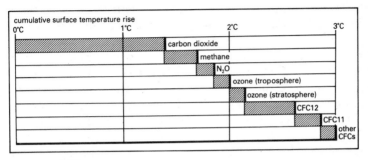

Source: UNEP

present trends continue, the Earth's atmosphere could contain another 100 parts per million of carbon dioxide by 2030. Even here, however, there are massive uncertainties. The warming which the greenhouse effect causes will itself reduce the need for fossil fuels to be burnt for heating, affect the biological productivity of plants outside the Earth's tropics, and alter the distribution of plant life and its ability to soak up carbon dioxide.

The plot thickens still further when we come to other greenhouse gases. Carbon dioxide accounts for about half the greenhouse effect but is a very feeble greenhouse gas indeed. Some of the others now being released into the atmosphere are up to a hundred times as powerful as carbon dioxide in terms of the greenhouse effect they can cause per molecule released. This means that carelessness about any of them can eat away seriously at savings in greenhouse emissions made by cutting down on fossil fuel consumption. Worse, some of them are chemically very stable. The various biological and chemical mechanisms acting to remove carbon dioxide from the atmosphere extract only one average molecule of the stuff in seven years, but greenhouse gases which are not part of mainstream atmospheric chemistry can remain in the atmosphere for over a century because there is no established process for removing them.

Of the alternative greenhouse gases, one which does have a role in nature is methane, the natural gas pumped from gasfields the world over and burnt in everything from gas stoves to steelmakers' furnaces. Burning it produces carbon dioxide, itself part of the greenhouse problem. But there are additional effects associated with methane itself.

Some of the small amount of methane in the atmosphere is artificial in that it has leaked from gas pipelines instead of being burnt as intended. Gas use is already outpacing growth in the use of other fossil fuels, because it is cheap to buy and easy to use. It also has friends in the greenhouse lobby because burning it produces less carbon dioxide for a given amount of energy release than burning oil or coal, because more hydrogen (which turns into water vapour) is burnt for every atom of carbon used.

Greenhouse gas studies published by the UN Environment Programme point out that records of methane concentration in the atmosphere run back only to the late 1960s, although ice core analyses like those used to determine carbon dioxide concentrations show a trend in line with more recent direct observations. Both indicate a steep increase in atmospheric methane at a rate of about 1.1 per cent a year, to a level of about 1.7 parts per million in the mid-1980s.

This increase is usually said in the literature to be roughly in line with world population growth, although it is by no means clear why the two should match with any particular precision. Methane is produced in the stomachs of ruminant animals and by some plants, including rice, and both these sources are increasing as the number of people wanting to eat them grows. (The ruminants are the origin of the newspaper story that farting cows are the cause of the greenhouse effect.) Some industrial activities, like coal mining, also produce the gas. As population grows, so might such activities, although some parts of the world economy are far more dependent on them than others. However, there are also powerful natural sources of methane which are less likely to

Figure 1.3 **Methane levels in the atmosphere**

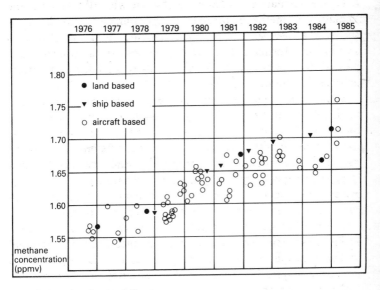

Source: UNEP

be affected by human activity, as we shall see. Even today's levels of methane in the atmosphere are only twice the concentration found in ice samples from before the Industrial Revolution.

According to UNEP, annual methane emissions to the atmosphere from all sources now equal about 425 million tonnes. But just as methane burns in a gas jet, so it is also used up by reacting with the oxygen in the air. An average methane atom probably lasts a decade in the atmosphere before being comsumed. The key point is that the Earth's capacity to absorb methane is falling behind human beings' capacity to emit it – especially with the help of all those cows. About 375 million tonnes a year seem to be consumed, and the growth of atmospheric methane is accounted for by the other 50 million tonnes a year building up in the atmosphere.

The rule of thumb that human population growth translates directly into growing levels of atmospheric methane means that projections of its future concentration depend directly upon expected population growth. Taking projections now regarded as central would imply methane levels reaching 2.3 parts per million by 2030 and 2.5 by 2050, half as much again as the present concentration and three times the pre-industrial level. This would imply a methane-generated greenhouse effect almost as large as the one now caused by carbon dioxide, since at the moment methane is estimated to cause about 30 per cent of the greenhouse effect observed, while carbon dioxide causes half.

However, the future for atmospheric methane is also bound up in the influence of the greenhouse effect on its production and consumption. Moist swampland which is home to methane-producing bacteria is the main natural source of the gas, although it is also produced from volcanoes and other natural sources. If global warming means a hotter, wetter world, it may also mean more places where such bacteria can thrive. One example is the Arctic, where global warming could turn large areas of low-lying tundra land into methane-pumping swamps. Moreover, the future of the atmosphere's ability to cleanse itself of methane is by no means clear-cut. The chemistry whereby methane reacts with oxygen is complex and can be slowed down or speeded up by the addition of trace gases to the atmosphere. In addition, it is possible to imagine breaking the link between population growth and methane production: if the developed world's decreasing taste for meat continues and spreads elsewhere, for example, the cows are likely to be in less demand rather than more.

Some of the same arguments apply to another of the greenhouse gases which occurs in nature as well as by human intervention. Nitrous oxide, one of the many gaseous oxides of nitrogen, is famous in the history of surgery as laughing gas, the first effective anaesthetic.

In nature, nitrous oxide is emitted to the atmosphere by the activity of soil bacteria. This natural process has been encouraged during the industrial era by the chemical industry's ability to produce cheap nitrogenous fertilizers. Adding these to the land increases its biological productivity, which is why farmers pay for them, but also increases the nitrous oxide they emit. This has probably added 10 per cent to the amount of nitrous oxide going into the atmosphere. The spread of agriculture into previously uncultivated areas like tropical forests adds yet more sources of the gas, and some industrial activities, like the burning of fossil fuels in vehicles and power stations, generate oxides of nitrogen from nitrogen in the fuel or in the atmosphere.

Nitrous oxide must be considered in the context of its immensely long residence in the atmosphere, an average per molecule of 170 years. According to UNEP, we could stop increasing emissions today and still have to live with accumulating atmospheric nitrous oxide for another two centuries. More realistically, according to UNEP, the likelihood is that the year 2030 could see nitrous oxide levels at about 0.375 parts per million of the atmosphere, up by a third from the pre-industrial era. Its present concentration is about 0.3 parts per million, enough to yield 6 per cent of the greenhouse effect.

The last group of greenhouse gases look in some ways like the most unforgiving. They are the CFCs, the chlorofluorocarbons, which occur in the atmosphere solely because people put them there, and some of their relatives like a group of chemicals called the halons. They contain the elements chlorine and fluorine along with carbon and hydrogen, and can last over a century in the atmosphere, but have attracted most attention in recent years because they are also implicated in a more immediate environmental threat than the greenhouse effect.

CFCs became the most notorious form of atmospheric pollution after it became apparent in the 1980s that they are responsible for environmental changes which could endanger one of the fundamental properties of the atmosphere, one which allows life as we know it to prosper on the surface of the

Earth. This is the ozone layer, a part of the upper atmosphere which filters out ultraviolet light from the Sun. This radiation is harmful to living tissue, and if allowed to reach the Earth's surface in unlimited amounts would cause alterations to living material including cancers and genetic defects.

Ozone is simply oxygen which is arranged in molecules of three atoms rather than the two in a normal oxygen molecule. The CFCs are lethal to it because after they are decomposed to release chlorine into the upper atmosphere, the chlorine reacts with ozone to destroy it. The severe damage to the ozone layer was discovered in the Antarctic and has also been reported from the Arctic.

CFCs are used for a range of tasks like removing heat from refrigerators, pushing materials out of aerosol sprays, as solvents and for making plastic foams. In the normal course of events, as UNEP points out, CFCs applied to almost all these uses find their way into the atmosphere – even fridges end up on the scrapheap sooner or later. Their allies, the halons, are used in fire extinguishers and for other industrial purposes.

The damage they might do to the ozone layer matters less than the fact that an average CFC molecule is 20,000 times as powerful a greenhouse agent as a molecule of carbon dioxide. This means that the CFCs were responsible for 14 per cent of the greenhouse effect observed in the 1980s, and because they last for over a century in the atmosphere the effect we see now will be long-lived even if radical action is taken immediately. The outbreak of allegedly ozone-friendly products in the supermarkets of Europe, Japan and North America after the revelations about the hole in the ozone layer, and international steps to scale back CFC production and capture CFCs instead of releasing them into the atmosphere, mean that the significant moves on CFCs are going to be taken because of the ozone problem, not because of the greenhouse effect. Until the mid-1980s, CFC production had grown at several per cent per annum, a rate which is set to go into reverse because of concern about the ozone layer.

However, the story would not be complete without mentioning another greenhouse gas: ozone. At a safe height (25–50km) above the ground, ozone is highly desirable. Lower down it is a pollutant, not least because it is highly poisonous. Its appearance in the lower atmosphere is mainly the result of inefficient burning of fossil fuels, which produces some ozone and also puts into the atmosphere gases like carbon monoxide which speed up the reactions which produce ozone in the Earth's lower atmosphere.

This means that the future picture of ozone in the Earth's atmosphere is a mystery which is unlikely to be resolved soon. In the lower atmosphere, ozone acts like a conventional greenhouse gas and for our purposes is a thoroughly bad thing. It has been increasing about 2 per cent a year and now accounts for about 12 per cent of the observed greenhouse effect. Its molecules are 2000 times as powerful as greenhouse gases as carbon dioxide. However, ozone has only a short residence time in the lower atmosphere before being converted back to normal oxygen. If the pollutants which cause it to appear were removed, it would be too.

On the other hand, the probable future for the ozone layer is of reductions in its thickness until the CFC problem is defeated. Reducing the scale of the ozone layer means allowing more infrared radiation through and might at first sight seem likely to add to global warming. However, the ozone layer does not remove the energy it absorbs from the equation. Instead it reradiates it into space and towards the Earth as infra-red radiation. Nobody knows whether the overall effect of a thinner ozone layer would be more greenhouse effect or less. In any case, as UNEP and others point out, future ozone concentrations in both the upper and the lower atmosphere depend on decisions about other emissions, so predicting them is inherently all but impossible.

The connection between the tiny traces of these natural and artificial gases in the Earth's atmosphere and the possibility of rising seas submerging islands, drowning coastlands and impoverishing low-lying countries is direct. The gases increase the amount of solar energy trapped on Earth instead of escaping to space; the trapped heat raises the temperature of the atmosphere; the warmer air melts ice at the Earth's polar regions and elsewhere: sea level rises.

However, the rate of exchange between greenhouse gases and rising sea levels is a more complex matter. There are three main problems. One is to calculate the amount of global warming which possible future levels of greenhouse gases might cause. After that, the connection has to be made between the warming and the amount of ice it might melt. The ice is not spread uniformly across the Earth. Instead, comparatively small areas, especially in the Antarctic, contain large amounts. A general rise in air temperature might not be reflected in the local weather in any particular area, so very large sea level rises might not occur even if the greenhouse effect does. And even then, the sea level rise which occurs will not affect all parts of the world equally. All over the world, islands and coastlines are dynamic systems, some rising from the sea and others sinking into it, for geological reasons unconnected to sea level rise. As we shall see, part of the problem is that even today, far more of the world's low-lying areas are endangered by rising sea levels than are experiencing rising land levels and retreating shorelines, a problem which the rising sea can only exacerbate.

However, in our present state of knowledge nobody really knows how much the Earth's temperature is likely to rise as a result of the greenhouse effect, much less how much sea level might rise as a result. Figure 1.4 shows calculations of the Earth's average temperature, and averages for the northern and southern hemispheres, since 1860. The overall picture seems clear enough – a warming by about half a degree in the last century. This increase does not sound very drastic, but it is enough to have significant climatic effects. It also parallels the growth in greenhouse gas concentrations which has galloped apace since the start of the Industrial Revolution. However, the change is trivial by comparison with the large falls in temperatures experienced during real ice ages or during the 'little ice age' of some 200 years ago. Paintings of that time show skaters and lunch parties on the frozen Thames in London. The scientific evidence from bubbles trapped in ice suggests that during genuine ice ages, the amount of greenhouse gas in the atmosphere is lower than it is today.

It is also important to realize that greenhouse gases do not act

Figure 1.4 Hemispheric and global mean surface temperature
changes, 1950–79

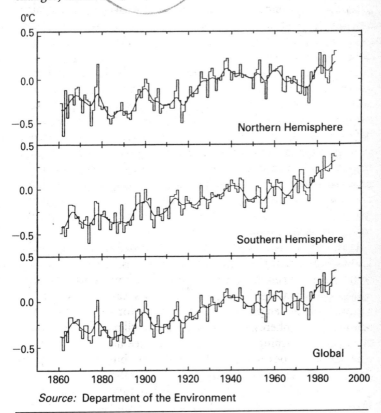

Source: Department of the Environment

instantly upon the atmosphere. It takes some years for a new
balance to be struck between heat retained in the atmosphere
and radiated into space as the concentration of greenhouse gases
varies, so that even if we stopped putting new greenhouse gases
into the atmosphere soon, the warming effect would continue
to build up. In addition, there are many complex pieces of

climatology which need to be taken into account in calculating the full greenhouse effect. For example, a warmer world will have a wetter atmosphere – as the temperature rises, so will water vapour concentrations. As water vapour is itself a greenhouse gas, the effect is self-reinforcing. More important for our present purpose is the effect of rising temperatures upon the Earth's snow and ice, mainly in the polar regions but also in mountain areas. Snow and ice are white, so they reflect solar energy which falls on them back into space – if greenhouse gases do not get in the way. However, melting them replaces them with bare land or ocean, which is bound to be darker and absorb more energy. Again, the effect is self-reinforcing because the heat retained can melt yet more snow.

It is possible, however, to build computer models – large complex ones – of the Earth's atmosphere which take all the known interactions into account, and to compare the results with experimental data on the Earth's temperature record. As Professor Tom Wigley of the University of East Anglia in the UK pointed out in a paper for the Prime Minister's climate change seminar, the models which have been developed so far do indicate a warming effect roughly comparable with the one observed, but it would be stretching a point to regard them as scientific proof of the effect. For example, no warming seems to show up in the data between 1940 and 1975, when the volume of greenhouse gases emitted to the atmosphere grew rapidly, but between 1910 and 1940 the warming effect was larger than the models would suggest. Since it is unlikely that the atmosphere would warm up in anticipation of future greenhouse emissions, this may simply be telling us that the Earth's climate has within it natural variation patterns of about the same size as the variations which the greenhouse effect might cause. A mathematical analysis by Wigley shows that a wide variety of effects seem to be at work to vary global temperatures, and that they operate on a number of timescales of up to a century in length. However, Wigley does not claim that the effects he has detected are large enough to explain the observed levels of global warming during the twentieth century, which he thinks are connected to the greenhouse effect.

Taking all these reservations into account, what do the models show? One sketched out by Wigley suggests that by the year 2030 the Earth could be 1.1 to 1.9°C warmer than it was in the mid-1980s. This warming would be added to the 0.5°C observed since the 1860s.

Jill Jaeger of the Beijer Institute in Sweden points out, in a paper produced for a 1988 meeting on the world security implications of climate change, that a massive array of forecasts of future temperature can emerge from the models now in use, ranging from some which provide only about 1°C of warming at the end of the next century to others in which temperatures are 5°C higher by 2040. Her middle view is compatible with Wigley's, but continues until the end of the twenty-first century and by that time has accounted for a 4°C temperature rise.

Fortunately for all concerned, the technology which is putting the greenhouse gases into the atmosphere is also providing the means to detect and describe the effect itself. In particular, the emergence of satellites for the observation of the Earth from space allows both more detailed measurements of the atmosphere and more refined data on the Earth's temperature. One example is a 1989 study by A. Raval and A. Ramanthan from the University of Chicago in search of the greenhouse effect. They chose to hunt it with a weapon called the Earth Radiation Budget Experiment, whose whole aim is to measure the flow of heat to and from the Earth by means of sensitive satellite-mounted scanners which can measure both infra-red and solar radiation to about 1 per cent accuracy. The study used this data for open ocean areas, which allow the simplest model-building and permit the use of surface temperature information gathered by the US National Meteorological Centre.

Their analysis suggests that the data strongly support models built to show the connection between the sea's surface temperature and the greenhouse effect caused by the increased amounts of water vapour produced as water temperatures rise. This would allow us to feel some confidence in models which imply an average temperature rise of perhaps 2–4°C in the coming fifty years. A few more years of observations would allow the effect to be seen in action with little possibility of error.

Nevertheless, the overall rates at which greenhouse gases are likely to build up in the atmosphere are still the source of considerable doubt. A 1990 paper by US scientists led by Pieter Tans of the University of Colorado, looking at just what happens to carbon dioxide in the Earth's atmosphere, makes the point. Tans and his co-authors point out that while the chemical processes which lead to the oceans absorbing carbon dioxide are familiar, the same cannot be said of the dynamics of ocean circulation and the biology of the oceans, both of which affect the actual rate at which the gas is absorbed. It is hard to obtain detailed information on the subject because the quantities involved are small by comparison with the amount of carbon dioxide in the oceans anyway. Adding half of the carbon dioxide emitted since 1850 by the burning of fossil fuels to the top 100m of the Earth's oceans would increase their inventory of inorganic carbon by just 1 per cent.

However, the authors' detailed models of the geography of fossil fuel use and oceanic take-up of carbon dioxide suggest that the oceans are absorbing a comparatively modest amount of carbon dioxide, perhaps only a billion tonnes a year. Subtract the carbon dioxide that remains in the atmosphere – up to 4 billion tonnes a year – and this leaves 2–2.7 billion tonnes to be lost elsewhere: in other words, on land. In practice this means in temperate latitudes in the northern hemisphere, near the major industrial centres where coal, oil and gas are burnt.

If this were not problematic enough, consider a 1990 paper by Cynthia Kuo and colleagues from AT&T Bell Labs in the USA, which points out that there seems to be a strong statistical link between the Hawaiian carbon dioxide measurements and the global warming of recent decades. In the same issue of the scientific journal *Nature* as Kuo's paper was a rebuttal from a scientist at the University of California pointing out that the temperature rises seem to occur in line with the carbon dioxide increases – but five months sooner, which ought to be grounds for suspicion. Another statistical analysis of global temperatures data since 1978, published in 1990 by scientists from the US National Aeronautics and Space Administration and the University of Alabama, finds no evidence for global warming at all. Their data cover the whole Earth – in contrast to temperatures from terrestrial thermometers, which cluster on land and near population centres. However, their results do not contradict the longer runs of data on global temperature rise over the last century.

As if this were not contradictory enough, it is even possible that fluctuations in the Earth's atmosphere, land and oceans are not the unique engine of climatic change. There is some evidence instead that they may be driven by a more fundamental force altogether – the raw amount of solar energy arriving at the Earth from the Sun.

The amount of solar energy delivered to the Earth by the Sun dwarfs all human energy production – the Crown Agents, part of the British government, once published a book on renewable energy sources called *Half an Hour a Year*, since it takes just thirty minutes for the Sun to deliver as much energy to the Earth as the whole human race uses annually. In more scientific terms, solar energy arrives at the Earth at a rate of 1.35 kilowatts – thousands of watts – for every square metre, at the top of the Earth's atmosphere before clouds and other obstructions get in the way. This is the average for the whole Earth from poles to equator over the whole year and is enough, for example, to power a dozen light bulbs per square metre of the Earth. By contrast, the amounts of 'greenhouse forcing' which atmospheric emissions might cause over the next fifty years will probably be only a few watts per square metre.

The Sun is a star, distinguished from all the others in the sky by being a lot closer to us than they are. Of the many types of star, some show radical variations in brightness of many hundreds per cent – enough to guarantee serious climate change problems for anyone living on a nearby planet. The Sun does not act like this, but its energy output does vary with time – for example, there is an eleven-year variation in the numbers of dark sunspots on its surface. Large numbers of sunspots are associated with a slightly lessened amount of solar radiation being emitted. It would take only about an 0.5 per cent reduction in solar energy output to account for the 'Little Ice Age' of the seventeenth century, which seems to have coincided with a period of low sunspot activity.

Work by scientists led by Richard Radick of the US Air Force Geophysics Laboratory suggests that stars like the Sun are capable of varying their energy output by up to a few per cent over periods of time of perhaps a decade, enough variation to account for serious climatic fluctuations on the Earth. An actual study of the solar energy arriving on the Earth between 1982 and 1988 shows that the variation during a single sunspot cycle seems to be about 0.1 per cent, probably too little to be a decisive cause of climate change. This variation would effect world average temperatures by some 0.02°C, well below the smallest change measurable on Earth.

A report produced in 1989 by the George C. Marshall Institute in Washington, which took the controversial stand that the greenhouse effect was not yet alarming enough to require action, made solar variability one plank of a platform from which it argued that the greenhouse problem contains too many unknowns for policies to be formed. According to critics of the Marshall stance, however, our knowledge of solar variability is so small that it may even be that solar energy output has been falling for the last century, reducing the greenhouse effect we observe. If so, the prospect would be even more dire than we now think if the process went into reverse and the Sun's output rose as emissions of greenhouse gases continued.

The real problem, however, is to find out just what sort of sea level rise might be associated with the global temperature rises of a few degrees in the next several decades now anticipated by most scientists: for example, the 1.1–1.9°C increase anticipated in Wigley's 1989 paper for the UK government.

There are at least two main ways in which sea levels rise in a warmer world. The first is the simplest: as the seas warm up (at least beyond 4°C, the temperature at which water is at its densest) they also expand – just as most things do when heated. Since the greenhouse effect does not also expand the planet itself, the jacket of water which covers most of it expands to cover a little more.

Next in the list of effects is the destruction of major ice masses at the Earth's poles. Here the problem is not sea ice floating in the ocean, as happens in the bulk of the Arctic icecap. If floating ice melts, it simply produces as much water as the volume it had previously taken up in the sea. The difficulty is ice which is not floating upon water: melting it adds to the total amount of sea, and therefore to sea level. This applies to melting small glaciers like those of the Alps or Scandinavia – for which there is evidence of rapid retreat in the last few decades – as much as it does to the massive ice masses like the Filchner and Ross ice shelves in the Antarctic or the Greenland icecap in the Arctic.

Johannes Oerlemans of the University of Utrecht in the Netherlands and the Alfred Wegener Polar Studies Institute in Bremerhaven, West Germany, has attempted to put numbers to these effects. Starting with the thermal expansion of sea water, he points out that it is already possible to account for part of the rising sea levels of the last century by reference to the greenhouse effect caused by air pollution over that period. From 1850 to 1980, global warming of about 0.7°C would give rise to a 6cm

Figure 1.5 **How greenhouse warming may affect sea level**

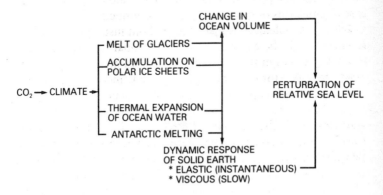

Source: Johannes Oerlemans

rise in sea levels. In future years, another similar rise by 2050 would mean a much larger rise of about another 12cm.

More complex is the melting of valley glaciers in the world's great mountain ranges. Accurate records of their history exist from many countries, and are all but unamimous that the glaciers are in retreat. Some seem to break the rule by putting on an unnaturally rapid downhill spurt, but even this is ominous – such behaviour is usually caused by warming which melts glacier ice to water and lubricates its flow downslope, and is a precursor of a severe retreat later. The total amount of ice in the world's glaciers outside Greenland and the Antarctic would raise sea levels by about half a metre were it all to melt. In practice, it will not. But Oerlemans reckons that enough of it could vanish by 2080 to produce a sea level rise of 11cm – give or take six.

The question of the major polar ice sheets is a more crucial and intractable one. They exist in a delicate state of balance whereby mass is lost by melting or by ice breaking off in the form of iceberg, but is also gained by falls of fresh snow. Greenland is a lot warmer than the Antarctic and loses ice by both means. A reasonable temperature rise there would put up world sea levels by 15cm. The surprise is the Antarctic. Here there is almost no melting because the temperature is so low. Global warming will allow snow to acumulate more readily, as will the increased water vapour content of the atmosphere. The total, amazingly enough, could be an icier Antarctic – icier enough to lower sea levels by 20cm by 2080, 5cm more than the loss of Greenland ice can raise them. However, there is enough uncertainty in the figures, even if the temperature increases are taken as given, for anything from a total sea level rise of 5cm to a fall of 15cm to be within the bounds of possibility.

No such positive benefits can accrue, however, from the possible catastrophic disintegration of the massive ice sheets of the Western Antarctic. These huge concentrations of ice include the Ronne and Filchner ice shelves, at the southern extremity of the Weddell Sea, and the Ross ice shelf at the south of the Ross Sea, the part of the southern ocean nearest to the South Pole. The ice in these shelves is kilometres thick. According to Oerlemans, the problem that arises is not simply that they

Figure 1.6 **Model projections of global warming**

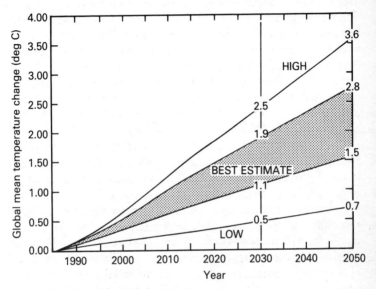

Source: Department of the Environment

might melt on a warm summer's day. Instead the issue is their relationship to the rest of the Antarctic continent. Warming the Antarctic weather would eat away at the front of the ice shelves, removing the points at which they are attached to the sea floor. At the moment the ice shelves exert many billions of tonnes of pressure on the glaciers of the Antarctic mainland. Once they became thin enough to lose their grip on the seabed, their ice

Figure 1.7 **Estimates of sea level rise during the next century**

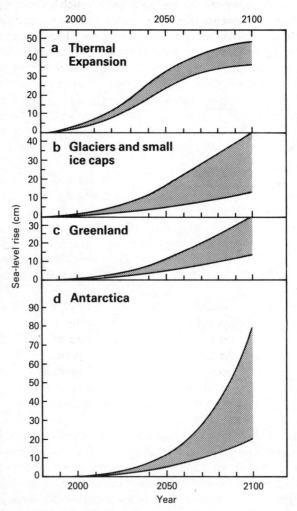

Source: Cities on Water Conference

would be pushed out to sea and that of the Antarctic continent itself would be able to flow north at a rapid rate which would cause it to melt.

There are many calculations of the effect such a disintegration might have. In essence they fall into two groups. One group says that the shelves will remain in something like their present form. This involves little sea level rise beyond that caused by limited melting at their northern edges. The other group involve severe break-up of the shelves: here the picture is naturally more stark. Typical figures imply that sea levels might rise by 1–2 metre in a decade. This effect might be long delayed, because it is widely thought that the West Antarctic ice sheets respond only slowly to changing temperatures, but if it happened it would do so comparatively quickly.

These considerations would support any calculation of future sea level rise from a few centimetres a decade to several tens of centimetres, with a probable maximum of perhaps half a metre by the middle of the twenty-first century. Perhaps the inaccuracies inherent in the problem are so large that there is no point in putting much more effort into the arithmetic, but more recent work has refined the possible dimensions of the problem a little.

For example, work discussed at the 1989 meeting of the American Geophysical Union in Los Angeles points to much smaller possible sea level rises than earlier research might suggest. Mark Meier of the University of Colorado pointed out that sea level rises have been measured at about 2.4mm a year for the last fifty years, or about 2mm a year more recently. Of this rise, the destruction of small glaciers (those not in Greenland or the Antarctic) accounts for about 0.5mm a year, and thermal expansion of sea water for perhaps 0.2mm. But the Antarctic and Greenland glaciers appear to be growing rather than shrinking, at a rate capable of producing over 1mm of sea level lowering a year. Meier is the first to admit that this points mainly to the complexity of the problem. As he says, the sea level rise so far seen is real enough and cannot be explained by the information we now have.

As Meier sees it, there is little scope for argument about one part of the possible rise in sea levels: the thermal expansion of seawater. If there is 3–5°C of global warming by the year 2050, 10–30cm of sea level rise has to follow. Lower down the agenda we come to small glaciers and the Greenland icecap, which he expects to chip in 16 (plus or minus 14) cm and 8 (plus or minus 12) cm of sea level rise respectively, for a total rise of 44cm. However, all these figures are dwarfed by his expected reduction in sea level rise as colder conditions in the Antarctic generate more snow and ice there. This, he thinks, could remove 10–50 cm from sea level. Although some of this would be counteracted by groundwater now trapped inside the Earth being removed to the oceans by higher temperatures, this calculation allows for a sea level change which could range from a 76cm rise to a 10cm fall.

The sea level rises discussed by Meier are lower than past expectations because of an increased belief in the capacity of Greenland and the Antarctic to soak up moisture from the atmosphere in the form of snow and ice. This could postpone major sea level rises for decades. However, Meier concedes that his models do not allow for a wholesale collapse of the major Antarctic ice sheets. He points to research which shows that there have already been major changes in the rates at which the Antarctic ice shelves melt and transmit water and ice to the ocean. Work published by UNEP implies that the polar regions will see much larger temperature rises than the world average, with rises of over 10°C in the Antarctic and the Arctic if the atmosphere's carbon dioxide content doubles, while some tropical regions might see temperature rises of less than 2°C.

A wide range of figures for possible sea level rises will continue to be offered by scientists working on a variety of different assumptions. In early 1990, Australian researchers suggested as a conservative estimate that a sea level rise of 15–75cm by 2050 might be regarded as reasonable. The real point to bear in mind is that the errors inherent in all these calculations are large, although a figure of several tens of centimetres in the next several decades appears in most of them.

But just what do these figures mean? A few centimetres or tens of centimetres sound like very small distances, but raising

Figure 1.8 **Changes in sea level over time**

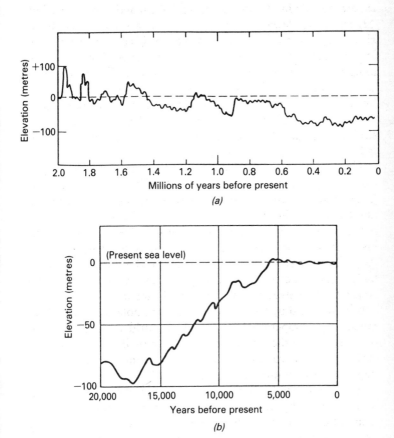

(a)

(b)

Source: Rhodes W. Fairbridge

the sea itself by such amounts is a severe prospect. As we shall see, it would mean erosion of coastlines, salt water onslaught on land and on water supplies, damage to cities, transport links and industrial plant, and further impoverishment for some of the world's poorest people – as well as an attack on affluent societies at vulnerable points. The next chapter looks at just how sea level rise might endanger people, natural resources, ways of life and economic activity. All the scientific evidence is that these problems are already with us – sea level is already rising – and are likely to increase.

Maldives

The Republic of the Maldives has an obsessive interest in rising sea level, as befits a nation whose highest point is little more than three metres above present-day sea level. Its government has taken a leading role in promoting world concern about sea level rise, especially within the Commonwealth, in the knowledge that even rises at the lower end of recent estimates would cause severe physical, social and economic damage to the country.

The Maldives – which lie south–west of India in the Indian Ocean – include about 1300 islands, 275 of which are inhabited and almost all of which are used economically for forestry or other purposes. Its population of about 180,000 (in 1985) is concentrated on twenty-five of these islands, including 46,000 on the capital island, Male[1]. According to a study for the Commonwealth Secretariat by Alasdair Edwards of the University of Newcastle upon Tyne in the UK, virtually the whole of the country's population, infrastructure and economy lies between 80cm and 2m above sea level.

However, the Maldives, in common with other low-lying tropical islands, do have one advantage over more northerly countries facing problems with rising sea levels – they can generate their own living sea defences. The islands are coral atolls defended from the sea by live coral, and if sea level rise does not occur at an excessive rate, coral growth may in principle be able to keep up with it. This pleasing prospect, however, is not a definite one. For example, rising sea temperatures may damage living coral by a process called coral bleaching. In addition, many of the

Figure 1.9 Maldives: Location of flooding and high wave impacts during June-July 1988

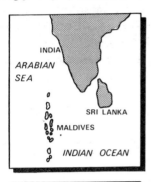

Maldive islands, according to work by scientists from the University of Wollongong in Australia, may be built of coral on a sandy base which will not follow sea level rises, rather than on the remains of old coral as occurs elsewhere in the world. Worse, it is possible that the increased erosion power of the rising sea will remove coral faster than it is capable of growing.

Rising sea levels pose threats to the Maldivian people and the staple areas of their economy. A recent growth point has been tourism, which is dependent upon the country's airport (1.2m above sea level) and upon developing uninhabited outlying atolls. Also at risk is agriculture. The cultivation of *taro*, one major crop, involves digging pits, sometimes below the existing high-tide level, while other crops like mango are highly sensitive to the increasing salinity attendant upon rising sea levels.

In addition, there is a severe risk to the islands in the recent policy of reclaiming land from the sea. Most of this land is less than 1m above sea level, making it exceptionally vulnerable to high tides and flooding. Worse, much of it has been reclaimed with the help of reef material, which would be more useful left in place. But an acute shortage of land for housing, especially on Male', means that there will continue to be pressure for more reclamation.

The size of the threat means that the Maldives government will be forced in future years to assess all its economic planning in terms of sea level rise, in particular plans for new sea works like jetties and causeways. In addition, industrial activities involving waste disposal to the sea which might damage coral growth will have to be restricted.

A particular problem is the islands' demand for fresh water, since fresh water reserves are likely to be invaded by saline water as sea level rises. The Newcastle study, however, regards this problem as trivial by comparison with the risk of saline invasion caused by the islands' increasing demand for water, a by-product of growing population and tourism. As more fresh water is pumped, the risk of salt water entering the aquifers from which it is drawn rises apace.

Dealing with problems like these will involve legislation to give existing government departments new powers and responsibilities – for example, to insist that industrial planning takes sea level rise into account and to ensure that prospective sea defences allow for future rather than present wave heights and sea levels. This will require the building of local expertise, and increased knowledge of the islands and their environment. This is true of all nations threatened by rising sea levels, but especially so in the case of the Maldives, a poor country where most house floors are about 30cm above sea level. Programmes to protect the islands from rising sea level or to modify housing and other buildings to cope with it would use up a significant share of the country's available investment funds. But even with sea level at its existing height, the islands are already under threat from flooding, salination and other problems which rising sea level will serve only to exacerbate. This means that the government of the Republic is one of the few in the world which is already being forced to take the issue seriously.

2 How the rising seas hurt

If an area – or a whole country – vanishes beneath the sea, anyone can appreciate the damage that has occurred. But even if sea level rise matches the most pessimistic of today's forecasts, the majority of its ill effects will not be so stark. Instead they will be subtle, gradual – and highly dangerous to human and other life, and to economic activity.

The most obvious and dramatic form of sea level rise is the severe damage which the most vulnerable areas – other than those which were submerged completely – would suffer from increasingly frequent flooding. The problem seems obvious, but in practice floods come in many guises. Several different types need to be taken into consideration in thinking about the damage which rising sea levels might cause.

The most familiar way in which sea level varies is by the tidal motion of the Earth's oceans, which are pulled, mainly by the Moon's gravitation, into a more or less elliptical bulge of water aligned with the Earth's equator. The precise mechanics of the tides have caused employment for the world's best applied mathematicians for centuries. The oceans are not simple, smooth test tanks filled with water: the coastlines which bound them are irregularly shaped and their seabeds contain shelves, mountain ranges and other irregularities. Particularly near major continents, there are usually large shallow sea areas where tides are especially complex. These forces mean that tides often give rise to strong currents instead of appearing simply as a general rise in sea level.

However, decades of painstaking collection of tidal data at coastal and offshore sites around the world have produced a mass of information which means that tide prediction is now a relatively exact science. Long runs of tidal data from St Helena in the south Atlantic and from Ireland in the north seem to show that the tides there have been essentially constant for centuries. The prospect now is that tides of a similar height to those encountered today will in future be overlaid on a steadily higher mean sea level.

The first problem with a change like this is that because tide levels are dictated by the exact configurations of coastlines and seabeds, and the depth of water covering them, it is not possible to add an average number of centimetres of sea level rise to the existing high tides and obtain an exact picture of the new tide regime at any particular site. The higher the tides and the more complex the topography, the harder it becomes to calculate the effect of a rise in tide levels. This means that the uncertainty is at its most intractable in areas like Bangladesh, where the tidal pattern is influenced by the complex outline of the Bay of Bengal – and where there is already one of the most perilous tide regimes in the world. The Bangladesh floods of 1970, which killed hundreds of thousands of people, were the effect of a 9m increase in water levels above the expected sea height.

In most areas of the world where flood damage is caused by rising sea levels the effects will be most apparent at extremely high tides, when new areas are most susceptible to flooding. The height of such tides will also determine the size of flood prevention works needed to keep them out, and so dictate the financial cost of keeping land available at a time of rising sea levels. But many of the most heavily populated coastal areas of the world are typically endangered not by predictable tides but by storm surges – in other words, by masses of water piled up by particular weather conditions. Many of the most dangerous floods in recent times – such as the January 1953 flooding of eastern England and the Netherlands – were caused by a combination of a high tide and severe weather forcing large volumes of water into a confined sea area.

According to work by David Pugh of the UK's Natural Environment Research Council, such storm surges have in recent historical times led to death tolls of up to 500,000. The biggest tragedies have occurred in Bangladesh, where a massive population lives on fertile sediment land close to sea level. But Pugh points out that there are several different types of storm surge in different parts of the world, each dangerous in its own way.

The most dramatic are the small, severe storm surges of the tropics and adjacent areas. They have a different name in every country where they are encountered: typhoons in Japan, hurricanes in the southern USA, cyclones in India, and so on. These are immensely damaging – and so compact that they tend to arrive and inflict their damage without meeting a reliable sea level recorder to tell the tale. If they do encounter a sea level gauge, the instrument is unlikely to survive. An exception is the case of Hurricane Elena, which hit Gulfport, Mississippi in the USA in 1985 and produced a water height reading of 1.5m above mean sea level. If the coastal defences of a particular area are designed to cope mainly with small tidal variations on mean sea level, rising sea levels will make the effects of such surges even more devastating than they are already.

The problem in non-tropical area is typified by the events of 1953 in the North Sea, when a depression moving to the south from Scotland to southern England piled water into the southern North Sea, from which it could escape only slowly through the narrows of the English Channel. To the surge was added a high – but not very high – spring tide. The surge in this case was some 2.6m. There have been at least six such cases this century on the coast of the North Sea, including one which caused severe damage in the German port of Hamburg in 1962. These events are rare but destructive. Rising sea levels will make them potentially far more damaging and as sea level rises, so will the size and cost of the coast defences needed to keep them at bay.

The same goes for tsunamis, the waves caused by the release of energy to the ocean by offshore earthquakes. In the Pacific, tsunamis (the word is Japanese) over 13m high have been recorded. They travel well: after the volcano Krakatoa in what is now Indonesia exploded in 1883, the tsunami was observed in the Atlantic. After the 1755 Lisbon earthquake, sea level at Newlyn in England "rose by 3m in a few minutes", according to Pugh.

Just how damaging these events will be in a world of rising sea levels depends crucially upon the tricky question of how the world's weather will be altered by global warming. Several studies have suggested that one effect of global warming will be an increase in storms and other forms of severe weather. As we have seen, global warming works by the build-up in the atmosphere of gases which prevent the escape of heat into space. This would reduce the cooling of landmasses at night, adding to the temperature differences between ocean and land which drive offshore wind regimes and perhaps making storms more frequent. Added to higher sea level, the effect would be more severe disasters occurring more often. However, even the most lugubrious schools of thought doubt whether increased atmospheric carbon dioxide will increase the rate at which severe earthquakes occur.

The first way in which tide and storms will affect coastal land will be an increased rate of attack by erosion. According to work carried out by Delft Hydraulics, a laboratory in the Netherlands, 70 per cent of the world's sandy beach areas have been undergoing net erosion in recent decades. Less than 10 per cent have been growing by the accumulation of new material, while the rest have been approximately stable. Beach formation is a dynamic process. As sea level rises, the response of most seafronts will be to form new beaches at a higher level: the coastline in areas which are not defended artificially will shift inland and a new, steep beach slope will appear and eventually stabilize. At the same time, the material removed from the coast will be dumped in deposition areas like tidal lagoons and basins, increasing the rate at which they fill up. Sea level rise will also disrupt the process known as longshore drift, whereby sediment moves along coastlines, by reducing the amount of sediment coming down rivers. More sediment will be deposited in river beds. Interrupting the drift process will reduce the ability of shorelines to resist erosion.

The Dutch scientists calculate that a 1m rise in sea levels would cause a 100–500m retreat of coastlines by mechanisms like these. Areas with special types of topography, like narrow ranges of dunes protecting large inland areas from inundation, could be even more severely affected. A study of the East and Gulf coasts of the USA finds that such a sea level rise would involve a $10–100 billion cost in replacement sand just to keep beaches stable: money that would be noticeable even in the context of the US defence budget.

Work done at the University of Maryland using old maps of the USA implies that along much of its length, the shoreline has been retreating at 30cm a year for the last century. On the Atlantic the rate is two to three times this, with a record of several metres a year set in Virginia. On the Gulf coast the record is some 5m a year of loss in Louisiana. The Pacific coast is more stable because it tends to be rocky rather that sandy, but even here some areas are affected by erosion, and the very narrowness of the beaches can mean catastrophic damage during severe winter weather. An example was the widespread destruction in Malibu, California, in the winter of 1981–2.

US researchers have also developed some methods for predicting erosion loss on the basis of projected sea level rise. The calculations result in a two- to five–fold increase in the present rates of erosion on the US coastline. This figure would be paralleled in many parts of the world where sandy coastlines face major oceans, but the figures would look less gloomy where there are rockier coasts or kinder seas. However, the loss of shorefront land by direct attack from the sea will probably be only the start of the story for the many countries whose coastal plains include large wetland areas. Intensive agriculture and industrialization have already caused these areas to be altered severely in recent years. Many wetlands in industrial nations, like the Somerset Levels in England, are now farmed, using methods which depend heavily upon continuous pumped drainage to prevent the land from reverting to its initial marshy condition.

Professor Stephen Leatherman of the University of Maryland in the USA has carried out a study of wetlands in the Eastern States of the USA. He claims that a historical sea level rise of about 15cm over the last century, increased by a factor of between two and ten by local subsidence in particular areas, allows us to find out about the problem directly instead of speculating about its possible course in future years. He focuses attention especially upon marshy coastland areas which, like beaches, are a delicate balancing act between land and water. A slow rise in sea level can sometimes be matched by the marsh itself as it builds up sediment and organic material like leaves and other plant remains. But if the sea level rises faster than the marsh can keep pace, it will be waterlogged and flooded.

A key example cited by Leatherman is the "backbarrier" marsh areas found in parts of the USA where an offshore barrier of some sort – typically a row of islands or dunes – protects the marsh from erosion. Areas like this rarely suffer from a lack of sediment for deposition; instead they are likely to be destroyed by the barriers themselves moving inland as sea level rises. This has already happened at Assateague Island in Maryland. In Virginia, South Carolina and Georgia, marshes are being destroyed after whole barrier islands protecting them from the Atlantic have been eroded away. The same thing could happen in Louisiana because of the destruction of coastal barriers which are now retreating at up to 10m a year.

The problem is less severe for the brackish marshes which form in major estuaries. Over the last century these marshes do not seem to have shrunk in the USA and elsewhere. This must mean that they are able to lay down enough river-borne sediment to cope with past rates of sea level rise, except where human intervention in the form of major engineering projects has deprived them of their flow of sediment. The immediate danger is also less, says Leatherman, for the "freshwater" marshes higher up river estuaries, although here rising sea levels will mean the intrusion of both salt water and salt-tolerant plants and animals into previously freshwater ecosystems.

But this picture would be radically worsened by a rise in sea level at the upper end of today's estimates. Adding several feet to present-day sea level would mean the formation of marsh ponds over large areas. These saltwater ponds would form and coalesce to produce large areas of open water in what today is marshland. At the Blackwater Wildlife Refuge in Maryland this process has already been in action. In the forty-one years from 1938 to 1979, 2000 hectares of the refuge, a third of its total area, were lost in this way.

Since rising sea levels will have their first effects on today's shoreline area, the forms of coast put at most severe risk will be those protected by unique plant and animal species which have evolved to cope with life there. A focus of special concern is the mangrove swamps which have formed in many tropical and subtropical parts of the world in tidal zones. These are a unique ecosystem in their own right, based on a number of species of the mangrove tree, which has evolved an ability – unique among trees – to cope with high concentrations of salt. Mangroves are also able to survive periodic immersion in water as tides rise and fall. The trees are mostly found above mean sea level – they do not tend to occur in places where they would be immersed in sea water most of the time – and below the level of the highest tides. Mangrove trees are superb machines for trapping sediment; this allows them to build up material above the low-water mark and provide sites for smaller plants and a wide variety of animals. According to Pugh, the animal life which takes advantage of the mangrove environment includes a range of species from crabs and barnacles to tree-living monkeys and unwelcoming creatures like snakes and alligators.

The popular image of mangrove swamps places them on the fringes of tropical islands, preferably in the Pacific. But they are also found on the coasts of major landmasses including the southern USA, home of the Florida Everglades, the world's most famous mangrove swamps. Large mangroves are found in Florida, the massive deltas of the Ganges and the Brahmaputra in Bangladesh; in river deltas in Australia's Queensland and Northern Territory; and in Sumatra, Papua New Guinea and smaller Pacific islands. There are extensive mangrove areas in West Africa and parts of South America. In India they are found in West Bengal, Tamil Nadu, Gujarat and the Andaman and Nicobar islands.

The whole basis of the mangrove ecosystem is that trees trap sediment and build up stable landmasses. This allows the energy of incoming waves to be absorbed, so that erosion in the swamp area is either very gradual or actually negative, with material being built up rather than lost. This means that the mangrove area itself is stable and the land behind it is protected from the sea. How will this apparently tough arrangement cope with rising sea levels?

The problem, according to work carried out for the Commonwealth Secretariat by Joanna Ellison of the University of California at Berkeley, is that a mangrove swamp is established only with difficulty and in specific surroundings. Individual mangrove trees can live on almost any tropical coast, but a complete mangrove swamp and all that goes with it calls for a comparatively calm sea, a gently sloping beach for trees to colonize, and a sea level stable enough for long enough to give the system a chance to get started. Ellison says that a look at the way in which mangroves have coped with past sea level change allows us to form an idea of their ability to cope with future rises. Ecologists studying particular mangrove swamps can form an idea of the past sea level changes they have experienced by mapping the layer of tide-carried material which builds up in the swamp within the top half of the range of tides. This means a picture can be built up of local sea level changes and the mangrove swamp's adaptations to them.

Thus work carried out on Grand Cayman in the Caribbean shows that during the great sea level rises which occurred between 6,000 and 10,000 years ago, mangrove swamps on the island remained small and restricted in area during the main period, when rises averaged some 9cm a century. This sea level rise was accompanied by a migration of the swamps inland as their previous sites became unusable. The same story can be read in the mangroves of Florida, where mangrove swamps became well established only between 3000 and 3500 years ago. This work has also been replicated, with variations due to differing local conditions, in Pacific islands like Tonga, and in Australia.

The point is that according to the research, the large mangrove swamps did not exist during this most recent geological period of sea level rise. Only when sea levels became stable were the mangroves able to get started in sheltered areas. And of course, the rates of sea level rise which Ellison mentions in her work were only about a centimetre a year, not large by comparison with the rates now being envisaged as a result of global warming.

The problem appears to be that a time of sea level rise is also a time of severe sea conditions, with violent storms battering coasts where mangroves might otherwise get established. Offshore reefs and other barriers to wave action also suffer from increased erosion and their effectiveness is reduced during time of rising sea level. For example, there are now extensive mangrove areas on the island of Tonga in the Pacific, but only one small and very sheltered area is known to have existed during the period of rapid sea level rise.

Mangroves seem to find it hard to cope with rapid sea level rise because it endangers their basic way of interacting with the watery environment around them: trapping sediment in their roots. If the sediment is washed away, the swamp cannot form. Instead of flourishing swamps the observer sees mangroves in "refuge mode": individual trees or thin, patchy areas in which they cannot live long enough or build up enough sediment to get a proper mangrove ecosystem going. Ominously, scientists working on the Caroline Islands in the Pacific have found that mangroves in just such a condition were found there during the sea level rises referred to above.

It is essential to find out whether the factors which prevent new mangrove areas forming during times of rapid sea level rise would also endanger the existing mangroves. This is a more complex problem, but Ellison answers it with reference to fieldwork on low islands like Grand Cayman and the Pacific island of Tongatapu, chosen because the picture there is not complicated by a flow of sediment from inland. The swamps on both islands have built up mangrove material at a rate of 8cm per century, indicating that they would just be able to keep up with a rate of sea level rise of about this magnitude. They can, with difficulty, cope with higher rates of perhaps 12cm a century. Even the higher figure is well below the range of sea level rise figures now being produced by studies of the greenhouse effect.

A wide range of factors complicate this picture. Much of the material which builds up in the swamps is plant matter derived from the trees themselves. At the moment, mangrove trees in swamps away from the equator are smaller, and produce less vegetable matter, than equatorial ones. It could be that global warming will make the non-tropical swamps more productive. On the other hand, global warming could intensify the pressure on mangrove swamps by increasing the number and severity of tropical storms. In addition, other human-induced activities like the use of power boats in mangrove areas, which washes sediment away from tree roots, contribute to the pressures on them. So does the practice of cutting mangroves for wood. However, it is possible to protect mangrove areas from exploitation by setting up national parks and similar regimes, as the Indian government's Department of Environment has shown. The long-term aim of the Indian approach is to enlarge as well as defend the mangrove areas.

According to Ellison, the main danger to mangroves is not in places like India but on low, flat islands where there are no inland sources of sediment to replace the material which might be lost as sea level rises, or to help the swamps shift upward as sea level does the same. If sea levels do rise at the high rates in some present-day forecasts, the prospect is for the swamps first to be transformed as more salt-resistant species come to dominate them instead of the mixture of species now found, and then to be overwhelmed by flooding as even these species fail to cope with higher water levels and rates of erosion.

A similar problem – of a vital protection system for tropical coasts which may or may not be able to cope with global warming – arises in the animal kingdom alongside the plant kingdom of the mangroves. The animals in question are corals, colony-forming creatures whose fossil history runs back deep into geological time.

There are corals which are capable of living in cold, deep water, but the most important corals are those which live in tropical seas near sea level. The animals themselves are soft-bodied, but they secrete the calcium-rich shells generally known as coral. The mass of material thus formed provides a home for a wide range of other species like sea anemones and starfish and a wide variety of fishes, which combine to make swimming on a coral reef a most memorable experience. The coral also has structural significance. In some places it forms the physical base of whole islands; in others it forms reefs which are important in protecting coasts from erosion.

The problem which sea level rise poses for particular important coral reefs is taken up in this book's case studies of places like the Great Barrier Reef, Kiribati and the Maldives, but the general issue serves to show just how knotty a problem sea level rise is. The precise effects depend upon just how many centimetres of rise occur, but also upon the rate at which it occurs and the collateral effects, like global warming and the increased incidence of severe weather, that might go with it.

Corals, in contrast to mangroves, are true marine creatures and cannot survive lengthy exposure to the open air, so they concentrate at and below sea level. This means that the risk to tropical corals is not that they will be drowned but that the water level will rise at a rate which will damage their ability to live. The problem is that corals contain algae – plant cells – within their living tissue, and these are essential to the animal's continued life. Like all plants, the algae need sunlight to carry out photosynthesis. While tropical corals can cope with water depths of up to 30m, they cannot build up new coral material at an indefinite rate. This means that they are likely to be outpaced by sea level rise if the rate is too rapid. From the point of view of the coral, sea level can go on rising indefinitely provided it does so only at a measured pace – although other species might find this approach less to their taste. The key problem is not the rate at which the fastest "branching" corals, which can accumulate new material at over 10cm a year, can grow, but the rate at which the whole reef, including the "massive" corals which provide a continuous infill of material to give the reef physical integrity, can develop.

Barbara Brown of the University of Newcastle upon Tyne in England says that even coral reefs dominated by branching coral would be unlikely to match the high rates of sea level rise in some modern predictions. Large flat reefs seem to be able to accumulate at some 7mm per year, in contrast to about 11mm per year of sea level rise in some forecasts. However, in Mexico a reef has been recorded which was able to put on a spurt of upward growth of some 12mm a year in response to sea level rise, after a time lag in which water deepened over the reef. Cases like these are known as "catch-up" reefs. Brown has also identified num erous cases of "keep-up" reefs which have kept pace with sea level rise but she points out that there are also "give-up" reefs. These, found in the Pacific, are reefs where coral growth has failed to keep up with sea level rise, so that the reef dies when the water depth over it becomes too great for the corals to survive. Which reefs will respond in which way is difficult to predict in advance, but reefs which have a supply of sediment to allow them to support a rapid growth of branching coral have an advantage over those which do not, favouring corals near large landmasses with major river systems to deliver sediment over those associated with flat ocean islands.

However, the picture is complicated by the fact that global warming will probably be only one of the factors affecting the future of coral growth in the world's tropical seas. Corals are a definitive example of the way in which sea level rise, global warming and wider aspects of climate change might interact in years to come.

Corals are warm-water creatures, but like other life forms have a preferred temperature at which they like to live. The corals' ideal temperature for growth, on the perfect coral planet, is a steady 29°C all year round throughout the surface waters. On Earth, the seasons mean that this ideal is never realized. The complicating factor, however, is that temperatures just 1–2°C above the optimum are fatal to the coral – or, to be precise, to the algae which live alongside them. At the present stage in the Earth's history this is not a significant problem: there are few sea areas where temperatures stay above 29°C for long. Global warming, however, could change all that.

Indeed, there have already been a number of cases of "coral bleaching" in which sea temperatures of over 29°C have been held responsible for killing the algae in the coral and endangering the life of the coral itself. Coral is a hardy creature which can survive several weeks of bleaching (so called because killing the algae also deprives the coral of its colour). However, reports of coral bleaching have intensified in recent years and are associated with indicators of climatic change.

According to Brown, a prime example is the 1983 episode in which an unusual diversion of El Nino, a seasonal summer wind of the eastern Pacific, led to warm water bleaching corals "at locations that spanned the Pacific from Panama to Indonesia". Brown points out that studies of that event show that the fast-growing branching corals are the type most likely to be affected by bleaching – no minor consideration when these are the very corals being relied upon to keep pace with sea level rise. After the bleaching, some reefs switched from a steady state to a net loss of material by erosion.

The key issue for Brown is that the corals in today's oceans "have evolved in conditions where the rate of temperature change never exceeded that predicted for the next forty years". Nobody knows how well they will be able to deal with the stress, but bleaching events to date seem to have the effect of severely damaging even the corals they do not kill. In the West Indies, where there have been a series of bleaching episodes – the most recent, in late 1989, being reported by Thomas Goreau of the University of the West Indies – bleached corals grow more slowly than undamaged ones and their ability to reproduce is impaired, both highly dangerous changes during an era of sea level rise.

Not all the evidence is gloomy. Researchers point to the fact that other species of animals and plants are able to develop "heat shock proteins" which allow them to adapt to higher temperatures. They add that in the northern Arabian Gulf, where there are lengthy periods of summer with sea temperatures of 32°C, reefs seem to prosper.

The picture is further complicated by two other aspects of global warming. One goes back to the basics of the problem, the rapid release of billions of tonnes of carbon dioxide into the atmosphere by the burning of fossil fuels. Since corals operate by turning carbon dioxide (and other material) into shells, an increased supply of the gas might seem like an ideal present to help them cope with the other problems of global warming. Unfortunately, most tropical waters already contain so much calcium carbonate that adding more atmospheric carbon dioxide is not likely to help.

More crucial is the damage that might be done to reefs if there is a serious increase in the frequency of tropical storms. Coral reefs can survive such storms; they even have a helpful role in spreading corals from place to place to increase diversity. But the example of Hurricane Allen, which struck the West Indies in 1980, implies that at a time when reefs are under stress anyway, storms simply add to the problems. Hurricane Allen did severe damage to the reef, which has still not recovered.

Nobody knows whether this combination of factors will prove fatal to the world's coral reefs. However, all concerned agree that the picture is not brightened by human attacks on coral reefs – including dredging on them, mining them for building materials and allowing excessive tourist access to them. Corals also need clear water and cannot tolerate pollution. Reefs seem to recover only very slowly, if at all, from mining and other use, and countries which want to keep them working properly have to develop policies for doing so. These can include complete bans on economic exploitation of reefs, or even more draconian ideas like deciding to sacrifice an already-damaged reef for building material as a way of avoiding attacks on others.

Similar scientific complications – and policy compromises – are implicit in other ways in which rising sea levels will attack ecosystems on land and at sea. A study of the possible effects of sea level rise, and global warming in general, upon the world's ecosystems by the International Union for the Conservation of Nature and Natural Resources (IUCN) makes the point that physical conditions like temperature are only some of the factors which determine biological success. Competition with other species is the key to success and failure, and it is impossible to predict these in advance. There are also big time lags in biological systems. Trees, for example, can live for centuries and, once established, can remain long after the local conditions which allowed them to get started have changed.

The IUCN points out that in general, a warmer Earth will offer more scope for species now confined to the tropics – provided that other conditions, like rising sea level, do not cancel out the effect or reverse it altogether. Even mangroves and corals might be able to spread polewards: a 1°C rise in global temperature will displace the present climate zones by 100–150km towards the Arctic and Antarctic.

The idea of a more tropical world may seem attractive, especially to readers in non-tropical countries who happen to be reading this in mid-winter. But as we have seen, the idea that global warming will simply make life a little nicer for most of the world's inhabitants is not supported by the evidence. For one thing, climate zones cannot just shift in a mechanical pattern. One key determinant of biological activity is the amount of daylight and the pattern of daylight hours throughout the year, and this will not be altered by the greenhouse effect. In addition, the shapes of landmasses, oceans and other features of the Earth's surface have strong local effects upon the weather which will determine just how the warming affects biological diversity. In most parts of the world the bulk of economic activity, including human habitation, agriculture and industry, occurs near sea level and the effects of rising levels are likely to be negative. The same would probably be true of the effects upon living species.

Rising sea levels will affect ecosystems on land in two main ways. The sea will inundate and otherwise damage low-lying land, but it will also cause the existing pressure on higher land to be intensified as fertile, low-lying land comes under attack. Just as a 1°C rise in average temperatures moves climate zones over 150km nearer to the poles, so it also pushes them over 100m further above sea level at any particular latitude. This means that farmers will be tempted to shift cultivation uphill to make up for land lost to sea level rise. The idea will have particularly strong appeal in countries like Guyana (see pp.124–5), where most activity is now concentrated in a narrow coastal plain and most of the country consists of uncultivated mountain areas. These areas are under pressure already in many parts of the world, and severe soil erosion is occurring as new land is brought into cultivation. At the same time, it is also becoming apparent that such areas contain unknown species and resources and have value on their own account, and value to the human race. Rising sea levels will simply increase the pressure on them – which is already excessive in too many cases. At the moment much of the most disturbing environmental damage is being done in the tropics, but predictions indicate that the effects of global warming will be at their most marked in the temperate and Arctic regions, and the effects of global warming may well be most severe on plants and animals there.

The effects of sea level rise will probably be at their most severe in the case of both animals and plants with highly restricted ways of life. Obvious examples include the large number of birds which have very specific migratory patterns, feeding and breeding grounds. Many such grounds are in estuaries near sea level and are already under threat from drainage schemes, industrial developments, plans for tidal barrages and other forms of human intervention. The species which depend on them would normally be vulnerable during even slow, natural periods of climate change, because of the difficulty they would have in replicating their existing way of life or finding a new one. However, a normal rise in sea level might tend to move breeding grounds from place to place, perhaps displacing a particular area of marsh inshore, so that there is at least the possibility of the birds moving as the habitat moves. In a world where most coastlines are owned and used by human beings, prime agricultural land is not going to be turned over to use by migratory birds – instead, flood walls and drainage will be adopted to reduce the loss of farmland and the birds will be squeezed out between the plough and the ocean.

However, one severe effect of sea level rise will be felt by humans, other animals, and plants alike. Sea water is salty, and increasing the level of the sea will mean an automatic increase in the amount of salt water in contact with fresh water on land. With a few conspicuous exceptions like mangroves – trees which, as we have seen, are adapted specifically to cope with salt water – most land species need fresh water: from streams, freshwater lakes or the rain.

Saline attack from rising sea levels will probably be at its most acute on small islands, whose freshwater supplies are concentrated in a lens-shaped zone beneath the island, fed by rainwater. The smaller the island, the smaller the lens and the more vulnerable it is to attack by higher sea levels. At the same time, a smaller lens is also more prone to attack by pollution or by excessive water use.

Colin Woodroffe of the University of Wollongong in Australia has described how the lenses form and operate. The basic force behind their existence is the slight difference in density between sea water (about 1.025 grammes per millilitre) and fresh water (very close to 1gm per ml). This allows the fresh water to float on the sea water. The detailed shape of the lens depends on the exact geology of the island, but in all cases the lens is lens-shaped both above and below sea level. The fresh water floats on the salt water just like a ship at sea, with the important difference that the density disparity between the two is much less. Because the density of salt water is one part in forty more than that of fresh water, the surface of the lens – the water table – should in theory extend one-fortieth as far above sea level as it does below, obeying Archimedes' Principle like all other floating objects.

This idealized picture is disturbed by the existence of the tides, whose up-and-down motion blurs the clear line which would otherwise exist between salt and fresh water. The end result is a lower water table than theory would predict, and a thick brackish layer between the saline and fresh water zones. In normal circumstances, this brackish layer makes a stable transition zone.

The freshwater lenses of tropical islands have attracted considerable study, with deep drilling being undertaken to obtain data and computer modelling of the lenses carried out to determine how they work. Even in a stable regime, it seems that an island less than 400m across is likely in practice to be too small to build up a lens, which means that such islands are impracticable as human habitations. There also has to be an adequate supply of rain to feed the lens from above.

Just how rising sea levels affect this picture is hard to predict. Many small islands are in any case rising or sinking at rates comparable to the sea level rises now under discussion for tectonic reasons, in ways which affect localized groups of islands or even single ones. However, Woodroffe points out that today's islands have mostly grown up with sea levels at about their present position at least for the last 3000 or so years, and the

Figure 2.1 **Three ways in which salt water can intrude into a freshwater lens**

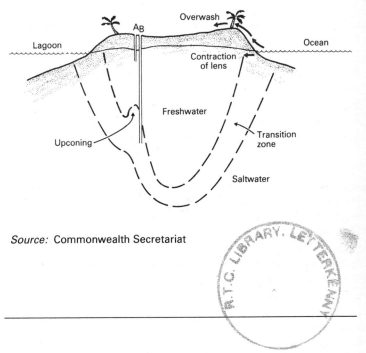

Source: Commonwealth Secretariat

variation in water lenses caused by the rise and fall of the tides indicates that they would be severely affected by changes in sea level.

The first effect of sea level rise on small islands is probably going to be a reduction in their size. This will reduce the water lens they support, and in some cases will shrink the island to below the minimum size needed to support one at all. The exact shrinkage depends upon the size of the island, its geology and the amount of rain falling on it. Since global warming is certain to mean changing weather patterns, the amount of available rain is likely to change as sea level rises. However, the key fact, which is not in dispute, is that the margin between an island with a satisfactory water table and one without is narrow. A drought which cuts off the replacement freshwater supply to the lens can allow rapid penetration of salt or brackish water to the roots of trees and crop plants, killing even big trees like coconuts very rapidly.

The same effect can be very quickly accelerated by pumping water from the lens at an unsustainable rate. Because of the way in which the fresh water floats upon the saline, lowering the overall water table by 2.5cm pulls up the bottom of the lens forty times as far – by a metre – making it increasingly likely over time that water from deep wells will become brackish and then saline. Salination of the lens would be reinforced during a period of rising sea level by the effect known as overwash or freeboard washover. During a severe flood which allows storm water to penetrate beyond the beach area and into inland zones, some of the water will percolate down into the water table and contaminate existing fresh water.

Just what this will mean for the living creatures on the island is difficult to predict in detail, but the outlines are clear enough. Many of the issues surrounding global warming and sea level rise involve effects which might, at least partly and in principle, cancel each other out. For saline intrusion there appear to be no possible benefits to counteract the probable damage. Almost all land plants, including the key food crops, are vulnerable to salt, with the exception of a few freaks like the mangroves. The same goes for virtually all land animals, including human beings. The intrusion of salt into the watermass used to support them will make islands uninhabitable to most of the species they now sustain. In the most extreme cases, human populations might become dependent upon rainwater and imported food. The islands' plant and animal life would be dominated by a small range of salt-tolerant species including types now found almost exclusively along coastlines. On some tropical islands, an important source of food is freshwater lakes which form in the centre, near which it is possible to grow root crops like taro.

These might be among the first areas to be attacked by salt water, since such lakes lie near sea level.

However, the exact reduction in freshwater lens size which would follow from sea level rise is impossible to predict and may in fact be zero. Rising sea levels would increase the saltwater attack on the freshwater lens. But because the fresh water floats on the salt water, a rising sea level would also push up the freshwater lens, causing it to expand into higher regions of the island. This might increase the volume of rock which can contain water.

A problem analogous to that of the freshwater lens occurs on coastlines on major landmasses, where a junction like the one around the lens is found between fresh and salt water. Higher sea levels will move the boundary landwards, making agriculture more and more difficult and endangering species which are not salt-tolerant. Added to the effects of increased erosion and storm damage, the overall attack on coastal habitation could be severe. In built-up areas, higher sea levels will mean salt water getting into storm drains and sewers and polluting freshwater supplies; this could necessitate resiting and rebuilding sewage farms, waterworks and other expensive capital plant. Also under attack will be major coastal engineering works like estuary bridges, which have been built to stand up to the forces placed on them by higher and stormier seas.

We can see some of the possible effects of salt penetration on undeveloped coastlines by looking at developments over the last few decades on the Atlantic coast of the USA, where extensive sea level rises have already been observed. According to Jim Titus, sea level rise project manager at the US Environmental Protection Agency, the saline intrusion caused by sea level rise has been worsened by droughts and by canal dredging and other human activities, allowing a clear view of the possible course of future events elsewhere in the world. In Louisiana, researchers have already observed swamps dominated by cypress trees being turned into open lakes as the salt kills the cypresses. The Delaware River Basin Commission has found that the droughts of the 1960s came close to allowing salt into the water supply of Philadelphia and surrounding towns. New reservoirs may be needed to avoid a situation in which limits on salt in drinking water would be regularly exceeded. A drought following a further sea level rise might push salt water up to 40km along the Delaware river system, although it would take a massive 2.5m sea level rise for the effect to reach this scale.

Sea level rise, it seems, poses a whole range of threats to human life and to natural ecosystems, especially when it is considered in tandem with global warming and the growing pressure on resources in many parts of the Third World. Just how the attack would affect human habitation and economic activity is the focus of the next chapter.

Australia and the Great Barrier Reef

Along some 500km of the east coast of Australia runs the Great Barrier Reef, one of the most remarkable intersections of land, sea and air in the world. Rising sea levels will mean change for the Reef, but in contrast to many other parts of the world we cannot assume that all change there is certain to be for the worse. Australian scientists have looked in detail at the behaviour of the Great Barrier Reef and other major reefs over recent geological time – approximately the last 20,000 years – in an attempt to determine how they have coped with the many changes in sea level. The overall picture is complex, but the main point is that coral has continued to grow and flourish throughout this period and has shown an ability to regenerate itself in large amounts when conditions turned favourable.

Of key importance for the sea level rise debate is the question of just how coral like that on the Great Barrier Reef can cope with steadily deepening water levels. In general the record of the last 8500 years suggests that the coral can grow sustainably at up to 8mm a year; and it can grow at up to 16mm a year by developing branching corals containing large amounts of void space. However, the rate of growth of dense coral capable of forming large structural masses is below 8mm a year, a figure adopted as a possible rate of sea level rise by some experts.

A rate of sea level rise which would put the Reef below water would not be fatal to it. The coral can grow at depths of 10m or more below the sea, a figure far in excess of the most pessimistic sea level rises now being discussed. In the past, according to the geological evidence, the Reef has flourished despite being underwater and has grown up to the surface in the form it now occupies only when sea level rise has stopped and the coral has been enabled to catch up. The likeliest future for the Reef at a time of rising sea level is not destruction but radical change. The effect on areas which are now exposed at sea level would be to rejuvenate them by restarting coral growth as water covered them.

The principal problem area in this picture involves the Cays, 310 vegetated or bare islands built up along the Reef, usually to a height of 2–3m above high water. The highest rises to 4.5m and three of the Cays are inhabited. The Cays are formed by the refraction and diffraction of waves building sediment above sea level. Their habitat is unique; so are the flora and fauna found on them. Rising sea levels would obviously threaten to flood these areas for more of the time than they are inundated at today's sea level, but the increased amount of wave action on the Reef might also increase the amount of sediment moved on the flat reef areas where the Cays form and allow them to build up fast enough to avoid destruction. Some Cays which are built up by cyclone action on the seaward edge of the Reef might be destroyed by storms at first, but later increase in number and size as the stormier weather made more eroded-away coral material available for their manufacture.

However, the effects of rising sea level on these unique environments is bound to be traumatic over the planning lifetimes of government policy. Over a period of decades the vegetated areas in the centres of the Cays would be flooded by the sea level rises now anticipated, and this would cause severe damage to the plants and animals which live there. Some species of turtles and seabirds use specific Cays as breeding grounds and would be severely affected by the destruction of habitats there – although Australian scientists comment that these species seem to have survived some 12,000 years of recent geological time when the Cays and exposed reef areas were probably absent.

The other unknown quantity in these calculations is the possible effect of "coral bleaching" on the Great Barrier Reef. The north of the Reef already has temperatures of some 30° C in summer, beyond which coral finds increasing difficulty in growing. Higher temperatures than this over large areas of the Reef would be bound to weaken its ability to deal with other problems ranging from rising sea level to marine pollution.

3 Third World damage

The next two chapters look at the problem of sea level rise for human life – one looks at the developed countries and one at the Third World. The problems which rich countries with major capital investments and wealthy cities in coastal areas face from sea level rise are quite different from those of poor countries whose land is mostly used for subsistence agriculture, or for cities where a substantial part of the population lives in poverty. For the rich, sea level rise means the risk of economic loss and disruption, but in the Third World the risk for many people is of death rather than bankruptcy. With this perspective, even the most affluent of readers will hardly object to starting the narrative with the Third World and its people.

Some Third World countries will probably be virtually immune to sea level rise, because they lack any significant amount of coast. Mountain countries like Nepal and Bhutan in the Himalayas, Lesotho in Africa and Colombia in Latin America can afford to adopt the same superior attitude to the problem as the rich Swiss – although other aspects of global warming are likely to be less congenial even for these countries. But most Third World countries do have coastlines and low-lying terrain, and as in developed world countries, their populations and their economic activity tend to concentrate there. Major Third World cities like Calcutta, Lagos, Rio, Cairo and Bangkok are all near sea level, as are massive food-producing areas like the deltas of the Nile, the Mekong and the Yangtze, along with the bulk of the Third World's industrial assets.

Such deltas are the sites where sea level rise will have its most dramatic effects – except in the small states whose whole existence is threatened by the rising sea. They consist of land a few metres at

Table 3.1 Ten countries most vulnerable to sea level rise

Countries	Population	Per Capita Income
	(million)	(dollars)
Bangladesh	114.7	160
Egypt	54.8	710
The Gambia	0.8	220
Indonesia	184.6	450
Maldives	0.2	300
Mozambique	15.2	150
Pakistan	110.4	350
Senegal	5.2	510
Surinam	0.4	2,360
Thailand	55.6	840

Sources: United Nations Environment Programme, *Criteria for Assessing Vulnerability to Sea-Level Rise: A Global Inventory to High Risk Areas* (Delft, Netherlands: Delft Hydraulics Laboratory, 1989); income and population data from Population Reference Bureau, *1989 World Population Data Sheet*, Washington, D.C., 1989.

most above sea level. Even with stable sea levels, they are flooded more or less often, typically every five years at least. Rising levels make the floods more frequent and widespread and endanger new areas. At the same time, the rivers feeding the delta are slowed up earlier and drop the sediment they are carrying more rapidly. This sediment raises the beds of the channels in which the river is running, making flooding even likelier.

Sequences of events like this have occurred many times in the history of the Earth. The difference today is human intervention, which alters the picture in two opposite ways. It speeds the rate of change, but because that change is inimical to human life, it also engenders a human response. The most obvious is to build sea walls, river embankments and other defences against the rising water. But Pier Vellinga of Delft Hydraulics, a Netherlands laboratory, warns that these measures, once begun, are likely to be hard to cope with:

the population in such [threatened] areas will not allow for a "natural" restoration of the equilibrium, but will tend to build and/or raise dikes, construct drainage works and adjust the infrastructural works to a higher sea level. Technically it is possible to build dikes, etc, but this requires economic, social and administrative conditions that can probably not be met in many cases.

From a researcher whose country has for centuries attached a high value to reclaiming land from the sea and defending its existing land area from flooding, this is a warning which should be taken seriously. Vellinga continues: "Once building dikes is started, higher dikes will be needed due to subsidence of the drained areas, due to a further heightening of the river bed and river water levels and due to a social and economic demand to increase the safety behind the dikes." This means that many Third World countries are facing a decision to take on a losing war against rising sea levels with money and other resources which they have not got enough of anyway, in the knowledge that at best the costs of fighting – and not winning – the war are going to increase into the indefinite future. As we have seen (p. 42), large areas of mangrove swamp protect many Third World coasts against attack from the sea. In a normal period of rising sea levels, the mangroves would probably shift inland in response to the pressure on their existing terrain from increased erosion and other consequences. But behind most mangrove areas today lie highly productive agricultural regions, including fish farms and rice paddies.

According to Dr Martin Holdgate, head of the International Union for the Conservation of Nature and Natural Resources, who chairs the Commonwealth Group of Experts on Climate Charge and Sea Level Rise, the size of this threat to Third World countries points to a basic imbalance in the way the world will deal with the global warming problem. He told the 1989 Commonwealth Small States Conference on Sea-level Rise that the transitional costs of slowing or halting the emissions which cause global warming will fall mainly on the developed world, where the emissions occur, but the economic cost of dealing with the consequences will fall mainly on the Third World. However, the same meeting heard that it may be possible to reduce the cost by concentrating efforts on cheaper options than massive sea defences, including the revival of old technologies for building and food production. These are often designed to cope with flooding instead of shutting it out. In the same way, it may be possible to channel flooding to encourage large areas of land near sea level to silt up rather than try to shut the sea out.

Actions like these, however, depend crucially upon Third World countries having enough time and information to develop coherent plans for dealing with rising sea levels. The rates of rise inherent in present estimates of the greenhouse effect imply that the Third World has anything up to fifty years to prepare for the worst effects. Alasdair Edwards of the University of Newcastle upon Tyne, in a paper written for the Commonwealth Secretariat points out that with some political will and some management foresight, this opens up a wide range of possibilities.

Edwards describes the obvious idea of shutting the rising sea out with barriers of one kind or another as "the instinctive response", but one which may provide only a temporary, and economically unattractive, respite. Because the local effects of a global rise on sea levels will depend on their tide regimes and on geological factors like natural subsidence and elevation of the land, it is also vital to collect actual data from tide gauges around the coastline in question as a basis for deciding between cheaper and dearer methods of dealing with sea level rise.

According to Patrick Holmes of Imperial College, London, sea level information is only one of the types of knowledge which need to be obtained by Third World countries wanting to cope with rising seas. There must be a set of "coastal management objectives" setting out just what the country in question wants to achieve in a war against rising sea levels. Since one of the policy options, as always, is to do nothing, it is worth thinking about areas of land which it might be easier to sacrifice than to try saving. In most cases, however, the production of a detailed register of the assets of coastal areas will reveal an overwhelming number of important items, from the national airport to the main suburbs, where such an option is probably not a realistic one.

In this case, the next essential is a detailed database on the coastal environment and the processes which go on there, looking at patterns of erosion and deposition and other issues. This allows a list of options and costs to be drawn up, and eventually leads to a programme of protective works and other tactics. The real point may be to ensure that as much as possible of the work of adapting to higher sea levels takes place within the context of the decisions which people and organizations take all the time, even in normal circumstances.

One example is the siting of major industrial plant. Most such plant is dependent in some way on the sea, for shipping materials in and out and to provide cooling water for chemical plant and power stations. Many such plants represent very large parts of the fixed assets of the countries where they are sited. They also last a long time, usually for decades. And a piece of equipment like an airport, an oil refinery or a mineral-processing plant is not designed to be picked up and put down elsewhere, even in response to as major a challenge as sea level rise. Even if they were, doing so would involve hardship for their workforces and the nexus of local suppliers and businesses dependent upon them.

This implies that in many cases the correct tactic for dealing with sea level rise at such facilities is to defend them, but to avoid the temptation to put the replacement alongside when the time comes to invest in it. Building the replacement inland and uphill would allow the government which does so (since most major Third World investments are influenced in location by official opinion) to look far-sighted, and might be politically popular if it were seen to spread economic activity and wealth more widely around the country.

The main exception to this rule is probably investment in ports and harbours, the one type of industrial plant which makes little sense unless it is within a reasonable distance of the sea. The basic element of even the simplest port, except where natural harbours are found, is a protective wall of some kind to allow ships to shelter in safety. This means that most harbours are protected to some extent from sea level rise, especially from storm and surge damage. The principal danger is that the entire protected area will be isolated by storm attacks upon the region around it, which will typically include a town and communications links to the hinterland. In addition, harbours on rivers will probably be affected by increased sediment deposition, which may require dredging to keep the harbour in business.

This leads directly to the more difficult question of just how to deal with rising sea levels as they affect major coastal cities. There are few cases in history of a major city being defended against persistent marine encroachment, with the exception of the cities of the Netherlands. By contrast, cases are legion of cities which have been lost to the sea – from Dunwich in medieval England onwards – and of cities like Venice which are under damaging attack from the sea. The effect of rising sea levels on Third World cities will differ in detail from case to case, but the extensive literature on quick and slow disasters in the Third World suggests strongly that the effects will mainly involve impoverishing and endangering the people least able to cope.

By analogy with existing problems of land use in the Third World, where the poorest people tend to live on unstable land with the least access to water and power supplies, sewage removal and other resources, it seems obvious to this author that the major sea level cities of the Third World are likely to develop "flood ghettos" of a particularly frightening nature. These will be characterized by frequent flooding, salt-polluted drinking water, inadequate or non-existent utilities, and uncertain property rights for the residents.

In less marginalized areas of major cities, the first effect of rising sea levels will be on the buildings the cities consist of and the services which make them work. Increased salinity and higher water tables will generate a wide range of problems for city areas. Salt will tend to get into water supplies and will also attack underground cables and pipework far more rapidly than fresh water. Higher water tables will also render basements of buildings uninhabitable and drive up the price of alternative property. The end result will be a strong economic incentive for anyone who does not have to be at sea level to move upward and away. This might free land for the flood ghettos just mentioned, but will also involve a confrontation with the Third World's problems of landlessness and the unequal distribution of land.

Even more than developed world cities, those in the Third World tend to grow in a sprawling and ill-planned manner. For their poorest inhabitants the main criterion is the availability of land, however unsuitable, and losing part of a city to rising sea levels will simply mean pushing such inhabitants into even less suitable places. The problem here is inequality, not rising sea levels, but it is exacerbated by the sea level rise issue.

Also of importance is the economic significance of sea level cities in the Third World. Most Third World countries have only a few cities, containing a large and growing proportion of their population. An attack on the economies of such cities is also an attack on the national economy. This applies with special force to capital cities because of the government departments and company head offices they house. It may be possible over a period of decades to solve this problem by a well-designed programme in which key activities like government are shifted to new locations away from sea level rise problems. Other major employers would tend to imitate such decisions once it was clear that the commercial momentum which usually leads organizations to cluster together was no longer the key factor. Done with skill, such moves would also serve to spread employment and wealth more widely – a useful move since Third World countries tend to have only a few genuine centres of economic activity.

Third World agricultural land affected by sea level rise is an even thornier issue because of the low value of much of this land and the lack of power of most of the people who occupy it. In the simplest case, the end result of sea level rise might be simply to hasten the flight from such land by its inhabitants, who would be driven to seek work in the sea-threatened cities instead. Third World farmers affected by low prices, high interest rates and the other problems which face them already will have their incentive to abandon the land increased yet further as their productivity and profitability are hit by rising water tables, salty groundwater and an increased frequency of flooding. Only industrial-grade farming operations on the developed world model, mostly producing cash crops, will be able to afford the pumped drainage and sea defences needed to keep their land free from flooding, and even land free from flooding, and even these would be an inadequate defence against salination. The overall result would be a further shift of power against small farmers and in favour of the large, capital-intensive, low-employment farming methods which are likely to replace them.

The practical effects of rising sea levels on poor people are well illustrated by reference to the varied countries around the shores of the Mediterranean, which touches some rich countries of southern Europe as well as poorer nations of North Africa and West Asia. Work carried out for the UN Environment Programme on sea level rise in the Mediterranean indicates that the effects of the rising sea are likely to be multiplied by the impact of increased precipitation in the area, with dramatic effects for the region.

The Mediterranean area has a long history of human habitation and an almost equally lengthy history of being examined by archaeologists and other scientists interested in tracing its human and physical development. As a result, detailed records of its sea level through time are available. They reveal that rises caused by the greenhouse effect, combined with changes due to geological movements, may mean sea level rises of 3–20mm a year for the next fifty years.

The problem is the immense range of the expected rise. At the lower figure, cheap measures will allow extensive areas to be protected. At the upper limit, areas of land which have held human populations for millennia are all but certain to vanish. The human-induced sea level rise is uncertain but is likely to occur, and will affect all parts of the Mediterranean in the same way, but it will probably be outweighed by geological movements which will vary markedly from place to place. The message applies equally in other parts of the Third World subject to rapid geological change, especially around the Pacific, in the West Indies and in countries like Indonesia.

The Mediterranean sea level changes of the future will be felt on shorelines which are already suffering severely from erosion and from flooding during high storms. Two major river deltas, one in the developed world and one in the Third, make the point that sea level rise is not choosy about whom it attacks. In Egypt the Nile delta is falling as the sediment deposited in it is compacted down, and because of geological subsidence. In Italy, the same factors are affecting the Po delta although of course, the personal and economic effects for Italians affected by the process are likely to be a lot less serious than for Egyptians in the same position.

The erosion of land around the Mediterranean is being hastened, too, by a variety of human activities. Pumping water for agricultural and industrial use increases the rate at which land subsides, since the water, once used, tends to run off to the sea instead of being returned underground. In some areas this factor alone is lowering land levels by centimetres a year. Dams built on major rivers catch sediment which would otherwise find its way to the coast, wrecking the dams as well as increasing erosion downstream. In the case of the Nile, even 20–30cm of sea level rise would be comparatively easy to deal with by comparison with these changes, which are exacerbated by the increasing use of coastal land for food production, housing and work. Half a metre of sea level rise, however, would be a more difficult problem.

Future sea level changes will occur at a time when the Mediterranean coastline is already under severe pressure. The littoral zones of the Mediterranean countries make up 17 per cent of their area but hold 133 million people: 37 per cent of their population. Of these people, over 80 million live in urban areas. Forecasts imply that by the year 2025 there could be 200–220 million people in the same area, over 150 million of them in urban areas. If by that time the average level of the Mediterranean is over a metre higher than today, the clear implication is that there will be a major problem for Mediterranean countries and cities and for the people in them. Indeed, 10–20 per cent of the littoral population of the Mediterranean area will be directly affected by sea level rise, with the others suffering indirect effects.

Researchers who spoke to a 1988 UNEP conference on the Mediterranean effects of climate change came to the stark conclusion that "the expenditures for the alleviation of climatic impact could be met, without major problems, by countries with higher national income. However, countries with lower national income may have great difficulty covering the cost of counteracting the negative impacts of climatic changes." In other words, the best policy for dealing with Mediterranean sea level rise is to make sure that you live on the north coast, not the southern or eastern one.

This point is illustrated by a study of the Tunisian coast, based on an assumed sea level rise of just 20cm by 2025 and a 1.5°C temperature rise. The area in question has already been affected by long-term geological changes and by canal construction, deforestation and changing agricultural practices. There are also plans for dam construction on local river systems which would have drastic effects on the local environment. The temperature rise and the loss of river sediment will mean the disappearance of major parts of the local environment including the Ichkeul National Park, a seabird breeding area which is regarded as a significant local asset. The birds now feed in freshwater lakes which will become arms of the sea, filled with salt water, and will lose their ability to support the freshwater plants which the major species of migratory birds eat. Some birds living there, including flamingoes and other waders, might feel some benefit, but the present character and role of the areas would be lost.

UNEP has also compared the possible sea level rise effects in the Mediterranean with the possible problems elsewhere in the Third World. Speakers from the Caribbean emphasized that the problem of the Mediterranean area will be reproduced for them in quite different form, mainly because Caribbean countries have a shorter and less intensive history of industrial development. This means that they have no major dam systems, major water-using industries or the same thirsty conurbations. As these accumulate – if they do – such developments will mean major changes to the Caribbean environment. However, it is possible to list the parts of this environment which are likely to be affected particularly severely by sea level rise. These include river deltas, and behind them wetlands, seagrass areas and other vulnerable ecosystems like coral and mangrove. Among human and economic assets at risk are tourism and major centres of population, both of which concentrate at sea level.

In this context, the Caribbean emerges as an area of the Third World like many another: full of economic potential which will be hard to achieve for any number of reasons, and made critically more difficult by the rising sea. The picture of disadvantaged nations and people being harmed further by sea level rises caused by the developed world's air pollution habits presents policy makers with the starkest possible moral issues – or should.

Bangladesh

Bangladesh is a country dominated and shaped by water. Although it contains some low hills, the bulk of it is low-lying alluvial land, either river terraces well below 50m above sea level, or flood plains of the Brahmaputra, the Ganges and other rivers rising in the Himalayas, the world's biggest mountain range. These large areas are often only a few metres above present-day sea level.

So a rising sea level poses particularly acute problems for Bangladesh. The difficulties are enhanced by the country's poverty – its people are almost the world's poorest – and by the fact that the country is already prone to destructive disasters. Work carried out by Bangladeshi researchers for the Commonwealth Secretariat points out that in the financial year 1988/89 alone the country suffered its severest flood of the twentieth century, a violent cyclone, a series of tornadoes – and a drought.

The Commonwealth study – hampered somewhat by the lack of a systematic topographical survey of Bangladesh – looked at the possible effects of one sea level rise scenario, involving a 90cm rise by the middle of the twenty-first century and a 10cm fall in the land level, caused by old sediment compacting as new material is deposited on top. This rise in sea level relative to the Bangladeshi landmass would cause the loss of 23,000 square kilometres of the country's land area. Some of its districts (local government areas) would vanish altogether. So would 14 per cent of the cropped area of the country and 29 per cent of its forests.

Naturally, the effects of this change on the people of Bangladesh would be disastrous. The population was 100 million in 1985, and of these about 10 million would be driven out of their homes by the change. They would be most unlikely to find new land to occupy in Bangladesh and would instead join the already massive migration to urban areas away from sea level, probably moving to cities like Dhaka. By the time the sea level rise occurs, the numbers involved will be much greater.

The catalogue of assets which would be destroyed by this sea level

Figure 3.1 **Bangladesh under threat. Even a 50 cm sea level rise would inundate large areas of Bangladesh. A 2.0–2.5 metre rise would reach nearly to the country's capital city.**

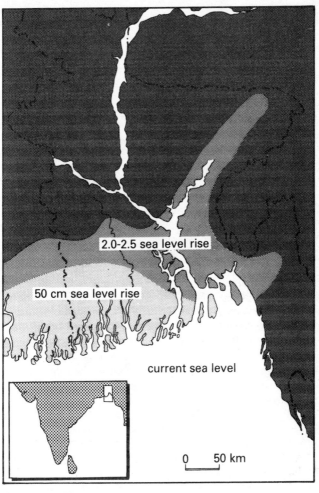

2.0-2.5 sea level rise

50 cm sea level rise

current sea level

0   50 km

Source: UNEP

rise is immense – the land in question houses over 14 million head of livestock, 100,000 households engaged in light industry, 8000 schools, 1500km of railways, 20,000km of roads and a wealth of other valuable facilities. But in addition, the rise would lead to frequent inundation in land at higher levels as water backed up into the river systems which dominate Bangladesh. Among the economic consequences would be the loss of the salt-sensitive rice crop grown in these areas. Rising sea levels and increased salinity also threaten 4000 square kilometres of mangrove swamps which are valuable both on their own account and for the protection against erosion which they offer.

Rising sea levels, and the greenhouse effect in general, would also increase the threat to Bangladesh from the floods and cyclones which already make life there dangerous for a large part of the population. The low land behind Bangladesh's estuaries is prone to flooding by water piled up by cyclones: the worst yet, in 1970, caused 300,000 deaths. Many of these deaths were of migrant workers bringing in crops, whose loss was a significant economic misfortune. To these floods in Bangladesh must be added river flooding and floods due to monsoons.

At the moment the only policy to ameliorate these potential problems seems to be the building of extensive flood protection works and barrages on the Bangladesh coast and along its major rivers. River flooding might also be reduced by flood control dams upstream in the foothills of the Himalayas. It would also be technically possible to reduce flooding by dredging the country's river channels, which are in constant danger of clogging up. But the cost of doing this would be massive, given that the task would have to be carried out virtually non-stop.

This means that a coherent flood control programme involving extensive civil engineering works is regarded by government and international aid agencies as the first tool needed to help Bangladesh cope with rising sea level. But the factors which make Bangladesh vulnerable to sea level rise, like its exposure to severe storms, are mostly beyond the country's control. Even the system of dams, locks and new canals thought up to deal with river flooding are really only a very expensive technical fix for the problem of soil erosion upstream. This means that international agreement on controlling the greenhouse effect and sea level rise, and on dealing with habitat destruction in the great Asian river systems, are the only way forward.

4   Rising seas and the rich

The countries of the developed world have in common an overwhelming control of the levers of the world economy. They contain most of its wealth, consumption and investment, its centres of credit, and of course its richest people. These facts all bear directly upon the problem of just how rising sea levels are likely to affect the developed world.

To start with, it means that rich countries can generally afford to take the steps which might be needed to deal with rising sea levels by a smaller diversion of national wealth and income than would be required in the case of a Third World nation. But in the case of sea level rise, the comparison is purely relative. Even in developed countries, rising sea levels at the upper end of present forecasts would damage large areas of land, attack major cities and industries, and otherwise cause damage which would be large even by the standard of the economies of the richest countries. After all, one of the main ways in which the developed world got so rich was its dominance of world trade. Most of this trade is carried out by sea, and the cities which grew rich from such trade cluster by the sea as well, along with the factories, refineries, transport systems and offices which go with them.

The wealth of the countries of the developed world is expressed primarily in these great cities, which with a few exceptions lie near sea level and near coastlines. While some developed world nations, like Germany, are mainly continental, few lack an extended coastline with ports and cities. So the attack by rising sea levels on the developed world will express itself in its most distinctive form in the shape of the attack on its cities.

According to a 1989 Venice conference on sea level rise and cities – the Cities on Water conference, of which more later – the problem is made more acute by the characteristics of modern cities and the areas around them. An example is the city of Hamburg in Germany, which despite being no less than 120km from the open North Sea is today under a serious threat of flooding. The problem has been growing because of the steady expansion of flood defences along the river Elbe, which connects the port of Hamburg to the sea. As the farmland and smaller towns between Hamburg and the sea have been protected by flood defences, the amount of tidewater coming upriver has increased steadily. As in many major river estuaries, the shape of the mouth of the Elbe funnels the incoming tide, making for water levels at high tide 70 to 120cm higher at the river than they are at the coast. In recent years the mean level of high tide has risen by about 10cm because of the flood defences downstream.

Now the flood defences of Hamburg itself are being raised by a twenty-five-year programme of capital spending. The city has already been attacked by the sea on several occasions, including a flood in 1962, which killed over 300 people, and another in 1976. The city has flood defences rising to a minimum of 7.5m above sea level, a figure which would be regarded literally as the height of extravagance in any Third World city threatened by the sea. Even so, it is not certain that these defences will be able to protect the city from the sea attacks of the future. The coastline is sinking under long-term geological forces, and if sea levels rise the problem will become correspondingly worse. In addition, Hamburg's determination to remain a major world seaport is incompatible with maximum protection from the sea. The dredging of the Elbe to allow the passage of large modern ships, especially container carriers, also allows the incoming tide a clearer run upstream, making for easier flooding.

The sheer scale of the economic problem which a threat like this means in a developed country is apparent from calculations carried out for Hamburg. Rising water levels are said to threaten 320,000 residents and 200 square kilometres of land: but the economic cost of such damage would be some $9 billion. At the same time, damage of this kind would also harm the whole economy of the Hamburg area and of Germany in general.

Hamburg is responding to the problem by spending another $2

billion-plus on flood defences. This option – which is open to any rich city in a rich country – means that other possible choices are neglected, including implementing a policy of shifting people and investment to higher ground and abandoning the threatened areas to the sea. In many cases this option is even more difficult for a large, developed world city than elsewhere because of the sheer economic and cultural value of the buildings and institutions near sea level. An example is London, where the Thames was the city's main motorway long before the invention of the motorcar. As a result, London's main public buildings, from the Bank of England to Parliament, Westminster Abbey and the Tower of London, are all on or near the river. So are the private institutions which go with them, from the company head offices with spectacular river views to the City of London itself, spoken of in the same breath as Wall Street as a centre of world finance.

Pressures to keep cities in existence rather move their activities and inhabitants elsewhere mean that the developed world response to rising sea levels will probably involve shutting out the sea rather than accepting its rise with good grace. London already has a flood barrier and other cities, like Leningrad, are getting them. In some cases, as on the Mersey in England, barriers being planned for electricity generating will also provide flood protection. But plans for anti-flood barriers are at their most complete in the world city most associated with the sea: Venice, a unique combination of water and land which, even before rising sea levels became an issue, was regarded as threatened by the waves.

The problem of Venice encapsulates in an exaggerated form the whole sea level problem as it affects the developed world and its cities. It can no more be moved to a new site than the Himalayas can: its buildings and people are what they are, where they are. And the sea itself is an inherent part of what the city is, admittedly on a more thorough scale than many cities manage. So the choice is to make a stand, or to remove the portable works of art and admit that the rest is not salvageable.

In addition, Venice is typical of many developed world cities in having problems with the sea which started even before rising sea levels became a world issue. One problem is that although

Venice is the unique relic of a medieval city state, it is surrounded by a large number of distinctly twentieth-century artefacts like chemical works and oil refineries, whose needs for pumped fresh water have caused land subsidence in the Venice area. David Pugh points out that Venice, like the Maldives, has no land more than 2m above today's mean sea level. The last fifty years have seen a 25cm rise in sea level relative to the city, as the soil under it has compacted and the water has been pumped out. This means that the future existence of Venice depends crucially upon artificial steps to defend it from the sea. The decision which has been taken is to protect the whole city by means of flood barriers, a step which will cost billions of dollars. This option is open to the city because of the riches which tourism brings to it. Indeed, central tourist areas of Venice are already subject to regular flooding at high tides. Even a few tens of centimetres of sea level rise would add to the seriousness of the situation.

The flood defences of Venice can pay for themselves in part by adding to the habitability of the city. At the moment the sea – or rather, the conspicuously polluted water-based liquid found between the buildings – has already made lower floors of buildings uninhabitable, driving up rents elsewhere and forcing Venetians to move away. Slowing this process would provide massive economic benefits.

According to work carried out by a scientific working party convened by the UN Environment Programme in 1988, the problems which Venice is experiencing are simply a more spectacular version of similar problems visible all along the northern reaches of the Adriatic. They point out that a variety of human activities – like water pumping, harbour dredging, land reclamation and even the construction of sea defences for limited areas – have added to erosion problems along the whole coast. Building flood barriers for Venice only makes things worse somewhere else – typically somewhere poorer.

The effect is to make the whole area outstandingly vulnerable to the threat of sea level rise. Add 10–20cm to sea level and the water levels in the coast's characteristic lagoons (including the one on which Venice sits) would be more susceptible to serious floods, while erosion on beaches and elsewhere would increase apace. Tidal flats and reed beds, used for fish production and as the habitat for a wide variety of wildlife, would be eroded, along with beaches. Later in the twenty-first century, when sea level could be half a metre or more higher than it is today, the effects would be even more drastic. As well as Venice itself coming under severe attack, the economy of the coastal area in general, especially its shipping trades, would suffer severely unless the infrastructure used to support it was rebuilt completely.

The UNEP experts suggest that a somewhat fatalistic management approach to the problem is probably the one to go for. Components of this policy might include abandoning agriculture on reclaimed land, allowing it instead to flood and return to being fishable lagoon water. Industries should move inland and the abandoned area could be turned into a buffer zone between sea and land. Such a policy might work, but implementing it in a real world of farmers and landowners would be a tough option for a government to pursue in the face of demands for sea defences for existing land and property. And even in this future, it would still be necessary to tackle industrial and agricultural water pollution in the area, which at the moment is posing a more severe problem to the local ecology than sea level rise.

The same problems arise in varying forms all around the shores of the Mediterranean. According to work by Spanish scientists, it is seen with startling severity in the delta of the Ebro, in northeastern Spain. They say that the delta grew outwards into the sea for many years until 1960, but has now stopped growing. Instead, the lobes of land which project into the sea from the northern and southern sides of the delta are being eroded. The problem is the damming of the river, which has removed over 98 per cent of the sediment from the water which used to go to build the delta. Dredging and other forms of river modification have added to the rate of erosion. The Spaniards' look at the possible future of the delta is based on an assumed 20cm rise in sea level – far from the top of some recent estimates – and an increase in temperatures of 1.5°C. This amount of sea level rise would cause less damage to the delta than human activities have already. But there would be more erosion and more flooding, and the spread of salt water into lagoons would mean the destruction of bird habitats and the plants on which the birds feed.

Some of the same factors arise in the South of France, where a French expert admits with disarming frankness that "the unattractiveness of the coast in previous centuries resulted in most of the large urban centres of the Gulf of Lyon [being] built away from the shore so that they are protected from any serious impact of a foreseeable rise of sea level." However, some human activities like swimming and unloading ships have a way of migrating to the seaside even if it is not the last word in physical beauty. This means that the narrow coastal strip of the Gulf of Lyon is a patchwork of crammed holiday and seaport developments, interspersed with low dunes and coastal ridges. Although rising sea levels have yet to present a problem here, the increasing frequency of severe storms already has. The authorities have reacted with a programme of capital spending designed to control erosion and longshore drift, the movement of sand and other material along the shoreline. This involves planting in the beach areas, setting up nets to prevent material migration, and other tactics including building artificial sand dunes and other structures to maintain a usable strip of beach. These measures owe little to concerns about rising sea level but will have the effect of simplify ing the fight against it, since they create a defended area of little direct economic value to absorb the threat of rising seas.

By contrast, other areas of the Mediterranean are ill-prepared to resist rising sea levels – and have a lot to lose if the sea rises in line with pessimistic forecasts. The northern Greek coast of the Gulf of Thermaikos appears to scientists who have studied it to be a case in point. They find that the coastline has little defence at the moment against sea levels 50cm or more higher than they are today. The result would be the flooding of a fast-developing industrial area and a severe threat to the million-person city of Thessaloniki itself. Large areas of land on the Gulf coast have been recovered from the sea and stand to be lost again, including land used for the lucrative tourist trade. The Greeks suggest a programme of heroic engineering works designed to cut off the city of Thessaloniki from the sea – except for controlled access for ships – as a long-term solution to the problem. This would involve a 4,5km barrier in water up to 27m deep, even at today's sea level. It might be an attractive option to the property-owners of the city, but would involve growing and continuing expense. It would also send flood water elsewhere along the gulf coast, causing damage in areas which might be financially less well able to deal with it.

The conclusion for the developed world shoreline of the Mediterranean is that large areas would be little affected by a sea level rise of several tens of centimetres in the coming decades, although a larger rise would mean severe problems. And there are so many major cities and industrial developments on those shores that there would still be many serious problems even with small rises in sea level; and so many fragile environments, especially brackish lagoons, that ecological problems will also multiply as the sea rises.

The same issues arise in even more baffling variety when attention turns to the most significant of the developed countries, the USA. The country's sheer volume of carbon dioxide and other greenhouse emissions may incline sceptics to think that Americans deserve all they get from the rising sea. In practice, of course, the people who lose out are unlikely to be the ones who cause the bulk of the problem.

In the USA, the sea level issue is complicated by the sheer variety of types of coastline, from Arctic to tropical and from lowland to mountainous. Even if they were to be attacked by a sea level rise of the same severity, each would respond differently. If one considers only the contiguous forty-eight states of the continental USA, a clear divide is apparent between the Pacific coast and the Atlantic and Gulf coasts. As a geologically active and young region, the Pacific coast consists mainly of cliffs and small beaches. The stabler and older Atlantic and Gulf coasts are flatter and sandier. This broad picture is further complicated by the shoreline and offshore topographies found around the coasts of the USA, including offshore islands and barriers. These provide protection against erosion (but not flooding) on shorelines where the general ground level can sometimes rise by only a few feet per mile inland.

However, it is not safe to assume that even cliff-bound shorelines are immune from sea level rise. Work done in areas ranging from the Great Lakes to the Pacific and New England implies that cliffs in some of these areas are already eroding rapidly, especially where they consist of unconsolidated sediment rather than solid rock. At Cape Cod in Massachusetts the cliffs are already receding at 70cm a year, a rate which would be added to considerably by higher water levels. In southern California, the cliffs retreat not at a steady pace but in large bites, mainly in response to severe storms and sea surges. As storm surges get higher and more frequent – as they would in a warmer world with higher sea levels – the cliffs' rate of retreat would increase. But in Maine and parts of the Pacific coast where the cliffs consist of solid crystalline rock, surveys of the whole historical record show no erosion at all.

By contrast, the US coastline also includes large areas of wetlands which are close to sea level and are bound to be seriously affected if it rises significantly. Work by James Titus and others at the US Environmental Protection Agency shows that the threat to these wetlands would be severe even if sea level rise were occurring without human intervention. Normally, raising the level of the sea creates wetlands where there was previously dry land. However, surveys of the US coast reveal that the area just above present-day sea level where new wetlands might be created is a lot smaller than the area of wetlands which stands to be submerged. Two case studies, in South Carolina and New Jersey, suggest that a 1.6m (five-foot) sea level rise would wipe out 80 per cent of wetlands in the two states.

The position worsens rapidly when the extensive human use of the US coastline is brought into the equation. Like people elsewhere in the world, Americans are not going to abandon homes and jobs at the first sign of rising seas. Instead, they will push for protection. A case study of Charleston, North Carolina, shows that sea defences for protected areas would mean the virtual elimination of local wetlands. In addition, human activities (other than the release of greenhouse gasses) are increasing the risk to the wetlands. At the moment the biggest problem is in the Mississippi delta, where about 125 square kilometres a year of wetland are being turned into open water by a combination of natural land subsidence and the effects of dredging and levee construction, which remove material which would otherwise go to maintain the delta.

The loss of these wetlands would be an ecological upset of the most dramatic kind for the USA. According to Titus, wetlands are the natural condition of the bulk of the US Atlantic and Gulf coast almost from Mexico to Canada. The biological productivity of the swamps and marshes which make up the wetlands far exceeds that of the nearby dry land. Many species of birds, alligators and turtles spend their entire lives there. Others, including birds ranging from eagles to sandpipers, visit as an essential part of their life. In recent years efforts to conserve them have gathered pace, after centuries in which it was regarded as acceptable to drain them, use them as rubbish tips or otherwise destroy them. Titus warns that the temptation now is to stop conserving them on the grounds that they will drown anyway, an instinct which must be resisted.

In addition, the US coastline is home to a massive range and number of human habitations and economic activity. US researchers have looked at a variety of possible sea level rise problems for human activity, and come up with a wide range of issues which need to be addressed. For example, many of the main airports, from Boston to Hawaii, have been built on land reclaimed from the sea. On a larger scale, there are hundreds of miles of riverside levees which have been built to control flooding and are already under attack from rising water levels. Recent years have already been marked by severe flooding caused by failures of levee systems. Significant rise in sea levels would oblige state water resources departments and others with responsibility for river management to think about heavy spending on higher levees.

Moreover, the USA appears, according to a 1985 report, to have no fewer than 102 seaports servicing ocean-going shipping. All of them, and smaller ports for local traffic, will be forced to respond if sea level rises significantly. Tides and currents will be altered as sea level changes, and new wave regimes would mean new mooring patterns and cargo-handling methods. According to US scientists, the obvious effect of sea level rise – deeper harbours and channels – would apply only temporarily and in a small number of cases. More common, especially in the longer

Figure 4.1 North Carolina Coast with 5 foot rise in mean sea level. Darkened areas are likely areas of inundation without protective counter-measures.

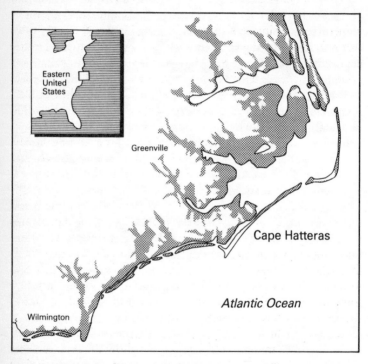

Source: Environmental Protection Agency

term, would be altered, and often intensified, sediment deposition, threatening the future of ports or causing an increased demand for dredging.

One way round this problem – already being tried at Brooklyn Naval Yard in New York, where land subsidence is built into the dockyard plan – is to site all new dock works at a safe height above sea level. However, piers, dock walls and wharves will be vulnerable to rising sea levels and to the increased frequency of storms which global warming involves, however heavily they are engineered. The amount of stress which waves impose on these structures depends upon the height of the waves, which increases as the water gets deeper, so that rising sea levels might call for a series of progressive strengthenings of sea defences. There would also be some less obvious threats. For example, brackish water harbours are not salty enough to support the marine boring worms which attack wooden structures underwater, but a higher sea level and more saltwater intrusion would allow them to eat away wholesale at such structures.

Shoreside structures like bridges, pipelines, roads, tunnels and railways are also all certain to have problems if the sea rises to meet them. In some cases the problem is in principle soluble – for example by replacing pipes more often as salt water corrodes them. Other issues, like the potential flooding of drains and sewers, may have to be tackled by major rebuilding. However, some major buildings and facilities which are found at coastal sites, from hotels to power stations, have a life expectancy of decades, which is compatible with the timescale for the anticipated rise in sea levels. This would allow the replacements to be sited and engineered for the sea levels anticipated on the late twenty-first century. Others, like offshore oil platforms, usually last twenty-five years at most. This means that they need not be rebuilt for higher sea levels, although their designers should be aware of the problem of increased stress caused by deeper water levels around them.

However, it is also worth remembering that even in the developed world's most congested areas, like the Atlantic coast of North-West Europe, not every square kilometre of coastal land is used for factories and houses, and the areas that remain undeveloped are of immense value ecologically and to local people. Work by British geographers shows that European coasts from Denmark to southern Spain, including the United Kingdom and Ireland, contain a vast array of sites of ecological value whose future in an era of rising sea levels needs a lot of thought.

Ted Hollis and colleagues at University College, London, point out that a nature reserve database, CORINE, kept by the European Commission contains 725 coastal sites. Estuaries, mud flats and salt marshes congregate in the UK, the Netherlands, Germany and Denmark, while sand dunes and sandy beaches tend to be found in Portugal, the Netherlands, Denmark and Germany. These sites vary widely in the amount of legal protection they enjoy. The UK, despite its mixed reputation in matters of environmental conscience, has designated most of its salt marshes as sites of special scientific interest, and the quasi-official National Trust controls over 750km of coastline to protect it from development.

The London geographers looked at the specification sheets for 178 Sites of Special Scientific Interest (SSSIs) on British coasts. An SSSI has no absolute protection against development, a sore point with British conservationists, but does suggest a place whose loss would be environmentally damaging. Of the sites examined, two-thirds are surrounded at least in part by built-up areas – except, of course, on the seaward side. Over a fifth adjoin urban areas on at least half of their landward boundaries. Only just over a third border open country on at least part of their boundaries, and only 6 per cent are surrounded completely by open land. These findings mean that one of the most treasured theories about sea level rise and wild areas turns out not to work. In a world without people – for example, during the sea level rises of the geological past – specific ecological habitats could shift inland in sympathy as sea levels rose. Marshes could move to higher land, estuaries could expand and sand dunes could retreat on to new territory inland. But these are not options which are open in a world of high land values where territory has owners and users. If there is a nature reserve between the sea and a housing estate, the inhabitants are going to respond

Figure 4.2 Marsh and mudflat areas of the UK under threat from sea level rise

to rising sea levels by asking for a sea wall between themselves and the water. They will probably feel a little sorry when the reserve vanishes, but are unlikely to volunteer to make room for it.

A detailed look at conservation sites makes it clear that some countries have a far bigger sea level rise problem than others. In Spain, for example, even the coastal conservation sites all run up to at least 40m above sea level, implying that only part will be

Figure 4.3 Cost of saving coastlines

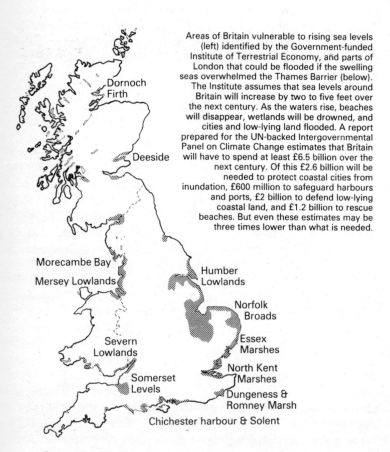

Areas of Britain vulnerable to rising sea levels (left) identified by the Government-funded Institute of Terrestrial Economy, and parts of London that could be flooded if the swelling seas overwhelmed the Thames Barrier (below). The Institute assumes that sea levels around Britain will increase by two to five feet over the next century. As the waters rise, beaches will disappear, wetlands will be drowned, and cities and low-lying land flooded. A report prepared for the UN-backed Intergovernmental Panel on Climate Change estimates that Britain will have to spend at least £6.5 billion over the next century. Of this £2.6 billion will be needed to protect coastal cities from inundation, £600 million to safeguard harbours and ports, £2 billion to defend low-lying coastal land, and £1.2 billion to rescue beaches. But even these estimates may be three times lower than what is needed.

Dornoch Firth

Deeside

Morecambe Bay

Mersey Lowlands

Humber Lowlands

Norfolk Broads

Severn Lowlands

Essex Marshes

North Kent Marshes

Somerset Levels

Dungeness & Romney Marsh

Chichester harbour & Solent

Source: The Observer

lost to sea level rise. The real problem is in countries like the UK or the Netherlands, where there are many sites whose maximum height above today's sea level is a metre or less. This implies that they could be totally lost under the kind of rises now anticipated by many forecasters. Many are swampy or lagoon habitats where the ecology, even today, allows for some flooding by the sea. These are particularly vulnerable, especially because many are in South-East England, where the land is already sinking below the sea because of long-term geological effects working through from the last ice age.

But the same research does find a considerable up side. Of the two basic responses to sea level rise – fighting and running away – it is possible that the second will be adopted in many cases even along the UK coast. This could mean the creation of new or replacement conservation areas. Most SSSIs border agricultural rather than built-up land. This means that a policy of retreating from agricultural land as sea levels rise could produce new land for ecologically valuable uses. A few years ago, such an idea would have been regarded as eccentric. However, farm policies of EC nations are already shifting as the European Community's Common Agricultural Policy, the main economic base of much EC food production, is reformed. In the UK, "set-aside" policies designed to take land out of agriculturally productive use are already being implemented. It could well be tactically desirable to set aside land threatened by sea level rise if the alternative is an expensive and ultimately doomed attempt to keep it in use, with the cost of defending it from the sea rising as salt intrusion and flooding lower its productivity. Overall, the researchers conclude, there are bound to be big habitat losses in sensitive areas adjacent to industries and towns, but the new habitat created by rising sea levels could exceed the area lost.

The point is emphasized by preliminary work carried out by scientists at the University of East Anglia and reported in 1990. Their own institution is at the epicentre of the problem. Large areas of East Anglia are protected from flooding by sea defences and lie below the levels of the highest tides. The cost of defending this acreage will rise with the seas. The same goes for

Figure 4.4 Louisiana Shoreline in the year 2030

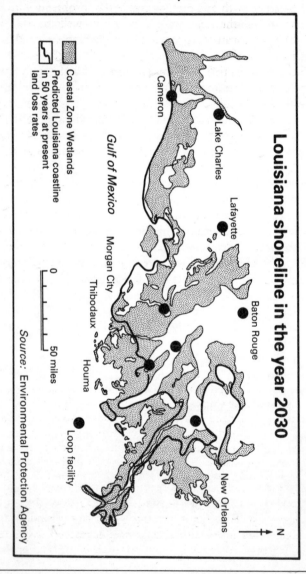

other low-lying areas of England like Holderness in Lincolnshire, which has already been subject to severe coastline erosion. At the same time, East Anglia has been the subject of rapid economic and population growth and already has severe problems with water supply. Most of the area's water comes from wells, and if salt gets into the groundwater there could be Californian-style restrictions on drinking water.

The message is that even in the comfortable developed world, people, economies and natural systems which are under stress anyway are all going to suffer even more seriously if the seas around their coasts rise significantly. And while some steps can be taken to reduce the impact, the correct policy choice may often be to realize that not everything that exists can or should be defended from the rising seas.

These problems are massive and real for developed world nations. But the countries in question at least have a choice – to channel resources into dealing with the problem, or to decide that it cannot be solved. Next we shall take a special look at some states to which even this choice must seem a luxury – the small states of the Third World whose very existence is in balance if sea level rises in line with present forecasts.

USA

Along the coastline of the USA, the sea laps against glacier ice and desert sand, and meets the world's most expensive urban real estate as well as desert land of next to no economic value. Sea level rise will affect all these types and uses of land, in a range of ways; here we look only at a couple of very specific sea level rise problems which are likely to affect fragile parts of the US coastline. In many parts of the world the extensive wet land areas between sea and land are of special concern because of their vulnerability to sea level rise. In the USA, two wetland areas chosen for particular study by the Environmental Protection Agency are those of New Jersey and in the region near Charleston, South Carolina.

A group from environmental consulting company Coastal Science and Engineering surveyed a dozen transects of wetland areas near

Figure 4.5 The Netherlands, showing areas protected by dykes and dunes. (White regions are more than 5m above sea level)

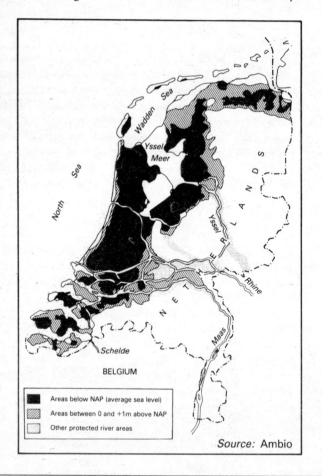

Source: Ambio

Charleston and looked at them in the context of sea level rises of 87cm and 159cm, and against the current trend of a 24cm rise because of general subsidence of the South Carolina coast, all by the year 2075. The region studied included wetlands ranging from those categorized perversely as "highland", high enough above sea level to avoid daily flooding, to others which are open water most of the time. The area is complex, with rivers, channels, small islands and other features which make a detailed analysis indispensable. The largest anticipated sea level rises could trigger severe ecological change in the kind of wetland which can survive. In this future, the area of the marsh underwater would rise from less than 30 per cent today to almost half. But highland marsh areas would be cut severely even if they were allowed to spread on to land which today is completely terrestrial.

There are a number of alternatives to this picture, involving various possible programmes of sea defences to preserve the marsh areas. For example, it is possible to protect the highland marshes – at the expense of eliminating some of the more frequently flooded types lower down, which would be squeezed out between sea defences and the rising sea. This research shows that even at the historically observed rate of sea level rise, the marsh area under study could shrink by 20–35 per cent in the coming century. Sea levels have been rising in the area by 30cm a century in the period for which records exist. The marshes do not mind this too much, but have no chance of coping with a rise of up to five feet which could result from a combination of sea level rise and land subsidence over the next century.

Applying the same methods to the New Jersey area yields results which differ in detail but emphasize the problems of coping with large sea level rises. Under the business-as-usual increase, the marshes can keep pace because new sediment can build up as the water rises. The marshes just get wetter and more frequently flooded. By contrast, big sea level rises can mean a loss of over half the marshland to open water. Even this picture requires the marshland to be allowed to extend freely inland as the sea rises, which is implausible because much of the coastline is already developed, or is likely to be in future decades.

Applying the same methodology to the USA suggests to the EPA that in an era when sea level rises and land is not available for wetland expansion, the USA could lose virtually all its wetlands by the end of the twenty-first century, and with them some its most physically beautiful, biologically diverse and ecologically productive terrain.

The Netherlands

Much of human history is bound up with the sea, but no country has been so shaped by its relationship with the water as the Netherlands. Large parts of the country have been created by reclamation from the sea, and almost half consists of land less than 5m above sea level. The Dutch are in greater peril from rising sea levels than the people of any other developed nation; but because of their history and their awareness of the problem, they are also more likely to do something about it.

Research by G. P. Hekstra of the Netherlands Directorate-General for Environmental Protection points out that the Netherlands coastline is protected from the sea in two different ways: half has sand dunes; the other half dykes. Each will respond differently to an increase in sea level, although both will be under increasing pressure as the sea rises.

Because the Netherlands exists only by dint of careful sea defences, the Dutch have adopted deliberate standards for flood protection, allowing for flooding at the highest surge expected every 10,000 years. This does not lead to a uniform height for defences, since the 10,000-year criterion is built into engineering calculations which also allow for local sea currents, the shape and type of dykes, and other factors. Usually the criteria are satisfied by building a dyke anything from four to six metres higher than the highest known storm surge from the area being protected.

Raising sea levels by 50cm will increase flood expectations in the low-lying areas of the Netherlands from once in 10,000 years to once in 2000 years, a fivefold growth in frequency. However, there are complicating factors: the increasing depth of the North Sea as sea level rises would reduce the risk of flooding – more volume for the water to be dispersed in – but increase the erosion of the country's flood defences. This effect has already been observed, along with an increase in the difference between high and low tides, which has had severe effects for coastal works and fisheries.

However, the real problem is that a third of the Netherlands is below today's sea level and another quarter is less than a metre above it. If sea level rises, this land will need to be defended more expensively than it is today. However, it will also be necessary to think about salt penetration beneath dunes and dykes into the country's prosperous agricultural land. The only way round this problem, according to Hekstra, is to raise the level of the entire Dutch water system – including lakes, rivers and artificial drainage – to match the rise in sea level. This would require a major rebuild of the country's sluice gates, dykes and other water engineering works.

Such a rise in sea level would also involve the Netherlands in a complete ecological reconsideration of the major rivers which carry most of the country's water. These rivers, the Rhine, the Maas and the Schelde, are used as drinking water supplies, for industrial and domestic waste disposal and for other purposes – in the case of the Rhine by users as far away as Switzerland. At the moment the pollution load on the rivers is high, but it can be carried away because there is a rough balance throughout the year between the water flowing into the Netherlands, in the form of rain and snow and via rivers, and out of it. Even now, however, there is a problem in summer when polluted river water has to be used for agricultural irrigation. This problem could become a lot worse in a warmer world because it seems likely that global warming would cause the winters to become wetter and the summers dryer, leading to serious pollution problems in summer. There would also be serious economic damage if the rivers were to become too shallow for navigation for part of the year.

Looking at a top-of-the range forecast of 1.5m of sea level rise, the Dutch point out that this would involve a major change to the country's way of life. It would be necessary to close the mouth of the Rhine, which now leads to Rotterdam, Europe's biggest port, and allow ships to gain access only via sluice gates. Water pumping would be needed in large areas of the country, and there would be major problems with salt penetration. Even for the Dutch, it would seem a little unfair to be flooded from the sea while watching the country's main rivers dry up for a major part of the year.

5 Small states and rising seas

The world contains many small nations, ranging from prosperous European micro-states like San Marino or Liechtenstein, which were not absorbed during the formation of the modern nations of Europe, to isolated island groups in the oceans of the Third World. These small nations generally have far narrower economies than larger countries, sometimes dominated by a single industry such as tourism. They also have much less geographical diversity than larger nations, which means that in an era of changing climates they have far fewer choices. If the world becomes dramatically colder, the inhabitants of Nepal cannot move somewhere warmer without abandoning their country altogether.

When small states face rising sea levels, they also face two special types of problem. First is the plight of a number of small countries whose whole landmass is so close to sea level that a rise of the size envisaged in today's forecasts endangers the actual existence of a whole nation. Second and more common is the danger that the economy and environment of a small nation, and the way of life of its people, will be placed under unbearable stress by rising sea levels and the associated physical effects.

Concern about both these issues has led many small states to take a strong interest in rising sea level – to such an extent that the Commonwealth Secretariat and the government of Australia supported a major conference in 1989 on the sea level rise problem as it might affect small countries. The venue, aptly enough, was Male', capital of the Maldives, one of the small countries in the front line if sea level rise becomes a severe problem whose alarm at the problem caused it to suggest the conference. Participants agreed that small states, because of their environmental, cultural and historical uniqueness, have an importance out of all proportion to their size and population. The question, in the case of small low-lying states, is how to ensure their survival in an era of sea level change.

The Maldives are regarded by experts as one of the world's nations most severely threatened by sea level rise, but a paper prepared for the Commonwealth Group of Experts on Climate Change and Sea Level Rise by British consultant James Lewis points out that this is only part of the story. He says that the small low-lying states where the threat of sea level rise is greatest are also countries which are in any case highly vulnerable to other types of disaster: floods and sea surges, heavy rains, high winds and other weather and tide-related incidents, and in many cases volcanoes, earthquakes, fires and droughts. He regards global warming as a continuation of present-day environmental hazards by other means. So these countries have got problems of their own before sea level rise gets started: but they also already have experience of dealing with environmental misfortune. This experience, he says, will be "the primary resource for addressing sea level rise and the problems and implications it brings" in small nations. By the standard of the problems they already face, says Lewis, the threatened rise in sea levels is "hazardous, but no more hazardous in most already hazardous situations than anything else", although he agrees that sea level rise might come as an unaccustomed shock to especially low-lying countries or to nations lucky enough to have little experience in dealing with natural hazards.

The problem for small island states is the sheer percentage of their area menaced by major sea level rise, and the correspondingly small amount of their land and resources which would be left intact. However, for island geographers small does not always mean flat. Some islands based on reefs and atolls are pretty well pure coastline, but islands of different origin, especially volcanic ones, can have the bulk of their landmass far above the sea. Lewis also points out that sea level rise is only one of the factors which alter island environments rapidly. We are used to thinking of the forces of nature as slow-acting, but earthquakes, floods and other natural events raise and lower land and alter geography overnight. In the same way, the human habitations which small islands provide are also not fixed in form: they change as the islands they occupy change. For example, the metre-per-century rise in sea level in the period from 18,000 to 6000 years before the present meant a vast reduction in the area of island habitats in the Pacific. As the sea rose, islands were split in two or removed altogether. Some were altered fundamentally by the change: cliffs which used to face the sea are now underwater and the island which remains, in the form of a low, flat area, is a mere stub of the previous larger and more varied one which is now mostly submerged.

Another problem which small countries face in dealing with sea level rise results from the fact that its worst effects are seen not in creeping inundation of land but in sudden floods caused by tide surges or severe storms. But while the world – quite correctly – notices a severe flood in a major country like Bangladesh, it is less likely to pay attention to a similar incident in an island state where the affected population numbers tens of thousands rather than millions. Lewis points to cyclones hitting small islands as a case in point. In recent decades hurricanes have demolished 80 per cent of the housing in Dominica, 95 per cent of the houses on Funafati in the Gilbert Islands and over half of the houses on Tonga, and done comparable damage to the crops of these islands. The most spectacular case was Hurricane Isaac in 1982, which caused widespread damage in Tonga, including the demolition of every house on the island of Mataku and permanent changes to the shape of the island of 'Uiha. These hurricanes, along with tsunamis, are among the most severe hazards to small, low-lying islands and are certain to be exacerbated by higher sea levels.

The obvious effects of sea level rise on small islands are to make them smaller still and to reduce their total coastline. Increasing the general level of the sea relative to the land will tend to marginalize land bit by bit, flooding it more often, causing salt damage to plants and allowing major storms more scope for inflicting damage like the destruction of coconut palms. The overall result will be to compress the area of the island available for activities like food production or tourism. This means that all small islands will feel increasing pressure on the coastal areas which already accommodate most people and economic activity. These are already the areas under the most pressure from population growth, economic growth, pollution and other pressures. But on islands with high land, the movement of population to coastal areas in search of work may be reversed as the coastal plain becomes steadily less habitable. The economic and social effects of such a transformation would be enormous for any Third World country, involving the reversal of all present trends. At the same time, inhabitants of low-lying islands and atolls far from the capitals of scattered island states will feel an even stronger temptation than they do already to move to the metropolis, whose leading economic and political role will probably be reflected in the money spent to protect it from the encroaching sea.

Lewis concludes that rising sea levels will probably be one strand in a web of "threat accretion" for small island nations, many of which, on the basis of the statistics, are good at protecting their citizens from the hazards they already face. For example, the citizens of most such states have far longer life expectancies than those of large African and Asian nations.

The small ocean states which are most vulnerable to the threat of the rising sea are mainly in the world's biggest ocean, the Pacific, which today has many hundreds of inhabited islands ranging in status from colonies to independent nations, and in scale from island states to groups which include dozens of inhabited islands.

The problem of sea level rise for the Pacific island countries is as diverse as they are themselves. John Pernetta of the University of Papua New Guinea points out that the land area of the Pacific islands is a mere 500,000 square kilometres, and that 85 per cent of this is in Papua New Guinea itself. Most such countries, even without the help of sea level rise, already consist mainly of sea. The Exclusive Economic Zones which most Pacific nations have declared around their land area consists of over 99.999 per cent water and under 0.001 per cent land, since a one-square-kilometre island around which an EEZ is declared can be the centre of a 125,000-square-kilometre EEZ. It follows that most of the Pacific island economies are dominated by activities associated with the ocean rather than the land, mainly fishing.

Pernetta, who chairs the Association of South Pacific Environmental Institutions, points out that the several types of Pacific island are all likely to be susceptible to sea level rise. Just having a substantial area of land a good way above sea level would not insulate an island from the problem. High volcanic islands, for example, have large fertile inland areas fed by copious rain, but are often fringed by coral and produce large amounts of food on low-lying land near the sea. There are also islands of mixed geology in which a volcanic core is surrounded by a low-lying area of limestone manufactured by corals – for example in New Guinea and New Caledonia. And there are at least two types of island of pure coral origin. Some are at or near present-day sea level and are based on living coral, while others have been raised above today's sea level by geological action. Islands like these are ecologically impoverished in terms of the range of species they support: for example, their tree cover is thin and they tend to be poorly supplied with fresh water. Low-lying atolls, in particular, are also highly vulnerable to extremes of weather, like hurricanes which can flatten everything on them, natural and artificial. Such islands, even today, "represent a marginal habitat for human existence".

It is also important to note that the islands of the Pacific, although small, are not sparsely inhabited. They have histories of human habitation running back thousands of years, and today have high population densities, which Pernetta puts at 386 people per square kilometre in the case of Nauru. One result has been heavy emigration: the world's biggest Polynesian centre of population is Auckland in New Zealand, and more of the citizens of Tokelau live in New Zealand than in Tokelau.

Also of importance is a comparable process of internal migration which is active in all Pacific nations. The capitals of most Pacific nations have high and growing population densities. Pernetta cites Majuro in the Marshall Islands (well over 2000 people per square kilometre) as the Pacific champion, although even Majuro is not in the same league as Male' in the Maldives, which has 56,000 people on an island 700m long and 700m wide: just over 47,000 per square kilometre.

The economic pressures which cause these population movements are the stuff of history all over the world: capital cities, for example, suck in both the educated and ambitious and the desperate and poor in every country. However, the issue arises in a particularly stark form in the Pacific islands because of their precarious economic circumstances and the severe pressures which are already affecting their unique cultures. The isolated and unusual characters of these islands are reflected in their human cultures – for example, on one reckoning a third of the world's languages are spoken in Melanesia. At the same time, most of the Pacific nations are heavily dependent upon aid and other non-commercial flows of money.

These problems are accompanied by ecological challenges which would be severe even if the sea level rise issue were not on the agenda. For example, the population of many islands and the sheer volume of material consumed by the people who live there have in many cases outrun any estimate of these islands' ability to cope. Many coral reefs near islands have been poisoned by sewage and other forms of waste, including toxic wastes like pesticides. In many small islands large items of scrap like old cars can be disposed of only by being thrown in to the sea.

Sea level rise will add to these problems in a number of ways. The most spectacular– complete inundation of land– will be important economically as well as geographically, simply because the population and agriculture of the Pacific islands are concentrated near sea level. On islands where the possibility exists, people will respond by shifting to higher ground. This will mean increased risks of land erosion as they grow food and build housing on sloping ground, and an increased attack upon the mountainous centres of islands.

At the same time, low-lying natural habitats which are already under pressure from human population growth will be subject to attack from a new direction as sea level rises. Pernetta points to animals like crocodiles and turtles – both already under threat in many areas – which need large areas of land near sea level to live and breed. If sea level rise causes islands to shrink and change shape to become steep-sided outcrops without any extended coastal plain, such species will lose out badly. In the same way, the increased flooding of land near sea level will affect many species in complex ways. Some saltwater crocodiles require flooding of nests – but not too much – to breed successfully and would be severely affected if floods became too frequent.

Problems like these are seen in an acute form in the Pacific in Kiribati, an array of thirty-three islands which straddle both the equator and the International Date Line. The islands are separated by hundreds of thousands of kilometres of open sea, and range in size from over 310 square kilometres (Kiritimati, much the largest) to under 5 square kilometres. The islands are all of pure coral origin and run to only a few metres above sea level except for one, Banaba, which has been raised to a maximum of 78m above sea level by geological forces. There are a total of 63,000 Kiribatians, living on islands with insufficient resources, mainly capable of producing only limited supplies of fresh water and a small range of subsistence crops including coconuts (which are also exported), taro and breadfruit. The subsistence economy of the bulk of the islands is replaced by a cash economy only in major towns.

Dealing with sea level rise in this economic and social context is bound to be difficult, although it will take decades for the full effects to become apparent, allowing some breathing space for policymakers. A study carried out by Lewis considers and rejects the idea of "citadels in the sea": islands reinforced by sea walls and built up above sea level by pumping material in from the sea floor. This idea might keep the islands in existence, but it would be massively expensive. Islands preserved in this artificial manner would also be unlikely to survive culturally, and because coral-derived rock is porous, sea water would rise in the centre of the island even if barriers kept it out at the periphery.

R.F. McLean of the Australian Defence Force Academy, in a study for the Commonwealth Group of Experts on Climate Change and Sea Level Rise, found that many of the Kiribati islands might be able to grow coral to match a reasonable sea level rise of, say, 30cm by the year 2030. However, the increased flooding and storms which will accompany this rise will be harder to resist. They will also be far more selective, attacking most severely the smaller islands and those which are orientated to expose most coastline to the major storm directions. This means that far more extensive study of the hazards to particular islands and coastlines will be needed to find out how to protect them from the effects of sea level rise. In a country whose government does not yet have the resources to tackle the problem, or even a section of a department responsible for the environment, the size of the challenge is clear.

The magnitude of the task is emphasized by a position paper produced for the Commonwealth Secretariat's 1989 sea level rise conference by the government of Fiji, whose Rural Development and Housing Ministry is in charge of considering the possible problem of sea level rise. The country of Fiji sounds idyllic from the Minister's description – set across the International Date Line in the South Pacific, with an area of 650,000 square kilometres of water and 18,000 square kilometres of land, and a population of 700,000 people on 300 temperate islands. The place seems to have just one drawback – it is in the Pacific hurricane zone, and has been struck by 150 more or less severe cyclones (although there is no such thing as a trivial cyclone) since 1875.

In the past the islands have been severely damaged by high hurricane winds, but have suffered even more severely from the storm surges which the hurricanes bring with them. A full-blown hurricane raises a 4m surge, and the biggest surges run up to 6m high. Both wind and surge damage are likely to increase if sea level rises. The surges which have been experienced in recent years have already caused sugar-cane-producing land to be taken out of use for up to three years at a time, and sugar is one of the country's main products.

Fijian calculations suggest that less than 3000 square kilometres of the country's total area is viable agricultural land. Of this, a sea level rise at the top end of the range now being discussed – about 1.5m – would be fatal for almost a fifth because of salt water intrusion into aquifers. The same rise would also remove the country's tourist hotels, which are built more or less on the beach, wiping out the other main pillar of the Fijian economy.

Most of the work carried out so far on sea level rise and small states has focused upon Pacific nations and upon the Maldives, for which there are shelves of reports from developed world aid agencies, local ministries and other bodies. The problem has been studied less adequately elsewhere. However, work carried out by the government of Barbados points to the universal nature of the issue. Leonard Nurse of Barbados's Coastal Conservation Project Unit points out that Barbados runs to just 430 square kilometres – big compared to the micro-islands of Kiribati but small when set alongside the major Caribbean islands like Jamaica or Puerto Rico.

This means that the Barbadian coastline is vital to the island as a whole. Ribbon development has enveloped large parts of the coast used for tourism and housing. Other parts contain a variety of preserved natural environments, including the island's last extensive mangroves. Industrial development is also concentrated along the coast and includes the power station, the oil refinery, the flour mill and the rum refinery. Nurse's inventory of the coastal activities and resources of the island runs the gamut from taxi stands to sea-egg harvesting.

Sea level rise affects this range of activities in a variety of ways – impeded, as so often, by the lack of any reliable data on just how large the effect is likely to be. The island is being pushed up at a rate of about 0.3mm every thousand years – not enough to be much help, but at least in the right direction. But even in recent years this effect has been overwhelmed by general sea level rise, which seems to be occurring at the rather more breakneck pace of about 0.6mm annually. This rate, of course, is the one observed before the sea level rises now on the agenda start to take effect. However, even a rise of this size could have what Nurse calls "a disproportionately great effect on the Barbados shoreline" because even today the island has only a narrow coastal plain. The highest point is 336m above sea level, so there is scant possibility of the whole place being inundated.

However, the topography of Barbados brings with it a separate threat. Because there are narrow beaches with gently sloping land behind them, even modest sea level rises could disrupt the island's most productive areas. This applies especially to the biggest foreign currency earner, tourism, which is almost entirely a coastal activity.

The disruption which rising sea levels might cause on Barbados would probably occur there, as elsewhere, in the form of spectacular flooding incidents rather than creeping rises in the general level of the sea. Barbados is sited in the track of the major Atlantic hurricanes, and it has been predicted that a major hurricane hitting the island at high tide would send 2m-high waves 80 metres inland at the south coast. There could also be significant slow-acting effects: for example, an attack by rising seas on the island's coral reefs. As we have seen, it is by no means certain that rising seas will kill off coral – although hotter seas might. But if they do suffer severely, so would the fish and other animals they support, and the tourist income they bring in. The corals dissipate energy from incoming waves and supply and capture sediment for the beaches, so that removing them would also reduce the island's ability to resist erosion.

In the case of Barbados, the steps being taken by the government to deal with the problem include getting proper instruments for sea level measurement and carrying out a proper survey of the island's coast and the potential hazards of rising sea levels. Elsewhere in the Caribbean, Trinidad and Tobago Environment Minister Lincoln Myers points out that the public education about the issue is also a priority, both to encourage debate and to prepare the people and governments for the need to take steps to deal with the issue. But he adds that replanting reefs and mangroves, measures designed to help build defences against the rising sea, can be started at once.

There are also signs that the small nations of the Third World are beginning to develop coherent ideas about tackling the problem. The 1989 sea level rise meeting in the Maldives adopted a final declaration which called for the small states which might be affected by sea level rise to collect information jointly, to participate in setting up a sea level monitoring system, and to propagandize worldwide on the possible plight of small states under threat from the rising sea. The small states also plan steps to protect their coastal lands and water supplies. But the Male' declaration also recognizes that the real action lies elsewhere – with the developed world countries and their emissions which are the cause of the problem. They also set up an action group, with Caribbean, Pacific, Indian Ocean and Mediterranean representatives, to follow up the decisions taken and co-ordinate approaches to the problem.

But the small states can also take action. The Fijian delegate at the Maldives conference pointed out that the rising sea levels and worsening weather which are likely to result from global warming are certain to mean more and worse disasters in vulnerable parts of the Third World. So the countries which are in the firing line (an apt metaphor since the speaker, Lieutenant Colonel Apolosi Biuvakaloloma, got the job by joining an army coup) need to put more resources into their planning for floods and storms. They must also think about coping with creeping disasters associated with floods and salt invasion. Biuvakaloloma draws encouragement from the way the Dutch have been able to create a prosperous country behind sea walls, and asks whether the Third World states facing sea level rise can follow their example.

The problem with this alternative is that the canals and sea defences of the Netherlands were built up over centuries, in the latter stages by a country of considerable wealth and population. The Association of South Pacific Environmental Institutions, in a 1989 pamphlet designed to bring the sea level issue to political attention in the area, points out that building comparable institutions in the Third World would be massively expensive. The problem would be increased if they had to be built from the resources of small nations over a period of decades rather than centuries, unless the rich countries of the developed world contributed at least some of the cost. Even less desirable is the prospect (pointed to by the Maltese position paper at the conference) of small, poor, states devoting large sums reacting to sea level rise before the science has developed enough for the effects to be fully known. In any case, simple geometry suggests that sea barriers are likely to work best where they are used to protect a straight area of coast. Putting them around an island means that the amount of land defended for a given investment drops.

At the same conference, James Titus of the US Environmental Protection Agency pointed to the Maldives as an example of a country whose conditions before sea level rise became an issue have encouraged vulnerability to the rising sea. The very fact that the islands of the Maldives are not regularly scoured by storms means that building has gone on there down to well under 50cm above sea level. Titus would like all means, from new building regulations to advertising and public education (aimed at tourists as well as local people), to be used to emphasize the seriousness of the situation. He also advises centring new development on the higher islands of the Maldives chain, although the choices in this respect are more limited there than almost anywhere else in the world.

However, this perception of the problem also emphasizes just how little is known about the potential sea level rise disaster elsewhere in the world. A paper from the Land Department of the Kingdom of Tonga agreed that one priority for the country was simply establishing which parts of it would be endangered by particular sea level rises from 2m down to 50cm. The issue here is not merely one of water, land and social disruption. As Jack Hopa, Environment Minister of Vanuatu, told the conference, a change like the displacement of fishermen and their families has the potential to spread economic and social damage throughout a small island nation heavily dependent upon the sea.

Titus added, however, that that the history of the USA suggests that areas tend to receive protection only after they have been flooded at least once, which is ominous for the chances of Third World countries looking to cope with rising seas. It is possible that the small states of the Third World, which are so much more endangered by rising sea levels, will act faster on the problem. But the pressures on all such states are already so immense that a new and major environmental hazard could overwhelm them economically and politically even if the countries themselves are not drowned.

Figure 5.1 **Islands Of The Tropical South Pacific**

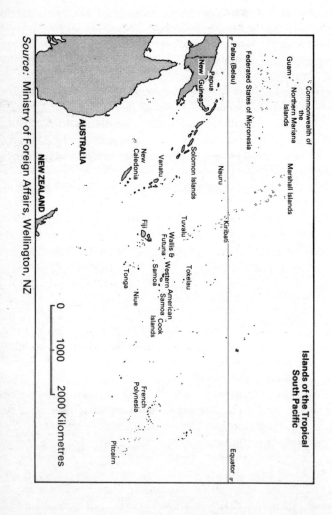

Source: Ministry of Foreign Affairs, Wellington, NZ

The Pacific

The islands of the Pacific are scattered, numerous and politically diverse. Their small populations ensure that they have little prominence on the world stage. But they face a common threat from rising sea levels, in common with developed and developing nations thoughout the world, especially other small island nations like the Maldives.

The South Pacific islands include twenty-five countries, covering half a million square kilometres of land in sixty times as much ocean. According to the Association of South Pacific Environmental Institutions in Port Moresby, Papua New Guinea, only seven of the islands are over 1000 square kilometres in size. They mostly have heavy population densities and depend predominantly on fishing rather than agriculture for their food. Many of the Pacific islands are low-lying coral atolls, although there are high volcanic islands as well as Papua New Guinea itself, which has high mountains.

The coral atoll groups like most of Kiribati, Tokelau and Tuvalu are the most vulnerable to sea level rise. A study carried out for the New Zealand Ministry of Foreign Affairs finds that under some of the higher sea level rise predictions now current it would be necessary to resettle the entire population of all three countries – about 70,000 people. "It does not appear feasible to adopt the kind of measures that could enable other countries with greater land and resources to adapt to the changing environment" of higher sea levels, say the New Zealand authors. This would involve the destruction of these countries and the forced removal of their populations.

The economic and social disruption which rising sea levels might cause in the higher islands would be scarcely less sweeping. Most of their key towns, including the capital cities of all the countries in the region, are coastal, and most of their agriculture is carried on at low altitudes. Important parts of Nuku'alofa, the capital of Tonga, would be flooded by a 50cm rise in sea level, which could occur in the middle of the twenty-first century. This would require industry, commerce and government to shift elsewhere at immense economic cost.

Since agriculture on the islands is mostly carried on at a low altitude above sea level, especially in river delta areas, rising sea levels could destroy the islands' ability to produce both cash crops and food for internal consumption. On Fiji, for example, sediment deposited by the sea could increase river flooding, whose effects would be added to those of sea floods and higher sea level generally. A solution to these problems will be beyond the Pacific islands' economic means, calling for assistance from developed nations.

The rising sea would be exacerbated by other weather effects associated with global warming. Of particular importance in the South Pacific is the incidence of severe storms, which could become even more frequent as the Earth's atmosphere becomes warmer and more humid. The most probable overall effect is an increase in cyclonic weather near the Solomon Islands, Papua New Guinea and even New Zealand itself. But some island groups, like Fiji and Tonga, may gain significantly from the new weather patterns as the cyclones move elsewhere.

However, the most insidious economic effects of rising sea level and global warming in the Pacific islands are bound to be offshore rather than on, since the area consists overwhelmingly of ocean, which provides its physical and economic base. The attacks on the islands themselves would probably be accompanied by shifting patterns of ocean currents which would affect the fisheries which give the islands their main income.

Rising sea levels would even threaten one of the islands' most fought-for assets: their 200-mile exclusive economic zones in which they claim economic control over surrounding sea areas. The exact drawing of such areas involves knotty problems of international law, but many of them are claimed on the basis of small low-lying islands. Their loss by erosion or flooding would remove them as a legal basis for the drawing up of such zones. This would turn large areas of sea from which fishing royalties can now be obtained – and conservation controls enforced – into free international waters, and also remove national claims for royalties if seabed mining in these areas ever becomes a reality.

6 Facing the rising seas

All serious environmental problems are routinely described as global, long-term, and indicative of a deep-seated contradiction between human activities and the natural world. In the case of sea level rise, the description is exact. The rising seas will touch every continent, and island countries which are part of no continent, and will go on doing so well into the lives of people not yet born. The problem arises because of natural long-term changes with timescales of thousands or millions of years, amplified by the worldwide effect of human releases to the atmosphere of gases which have no natural existence there, along with vast human-derived increases in concentrations of another gas, carbon dioxide, caused by consuming in a matter of centuries fossil fuels which have accumulated over hundreds of millions of years.

Dealing with the rising seas will call for new scientific knowledge, new engineering and technology, and new practices and methods in every field from agriculture to town planning. This chapter looks at some of the issues and their possible solutions; the next and last chapter will examine the policy responses which rising sea levels will require from governments and organizations in rich and poor countries around the world.

Sea levels, as we have seen, change over long periods as the Earth's climate varies. In the past, they have changed by massively more than even the most frightening rises anticipated by today's scientists. In the long term, trying to prevent such natural sea level changes would be about as sensible as passing laws against earthquakes. However, there is no doubt that something can be done about the artificially induced rise in sea levels which the greenhouse gases now being pushed into the atmosphere seem likely to cause. But anything which is done about the issue will not be intended primarily to prevent increases in sea level. Instead such measures will be part of a wider batch of measures designed to deal with the effects of global warming in general, from the displacement of crop-growing zones to the spread of tropical diseases to today's temperate regions.

The key to the conundrum is to find out as much as possible about the greenhouse effect itself, especially the rate at which it is affecting the Earth and the speed at which preventive measures might alleviate the problem. Here one arrives at the first piece of good news: the science-driven industrial growth which is at the base of the problem has been accompanied by the advance of the knowledge and methods needed to get its measure: instruments capable of detecting trace quantities of gases in the atmosphere, and computer systems which can use data about the atmosphere and the oceans to produce detailed models of the Earth's energy balance and the way it changes with time. The establishment of new organizations and scientific initiatives to investigate climatic change is one symptom of the emerging priority given to scientific knowledge on the subject, although the importance of the issue means that this is a case in which preventive action may have to be taken slightly ahead of definitive knowledge.

Another symptom is the array of major projects for gathering data about the oceans and the atmosphere. Many such projects involve satellites, the technology of choice for gathering large amounts of data fast about the Earth and its surroundings. Examples include the European ERS-1 and ERS-2 satellites, designed to produce data on oceans, ice cover and the atmosphere; the US NOAA weather satellites, to which sensors for greenhouse gases are being added; and the series of "polar platform" satellites being developed in Japan, the USA and Europe for earth observations in the late 1990s. The payload of these satellites is certain to include a hefty array of instruments for observing the Earth's oceans and atmosphere, yielding a deluge of information on everything from wave heights (vital for discovering whether storms really are becoming more common) to ice cover and even the Earth's temperature itself. Not long ago, the flood of data which these instruments will produce would itself have been almost impossible to deal with, but computers now becoming available are capable for the first time of coping with it and using it to develop realistic models of the physics and chemistry of the Earth's atmosphere.

The first issue which such new instruments and the computer models which they feed might set out to solve is whether the greenhouse effect exists. Although most scientific opinion is unanimous about the nature of the problem, some researchers believe that artificially induced greenhouse warming may not occur, may be minor by comparison with natural climatic change, or may even not be undesirable. The most distinguished of the sceptics is Richard Lindzen, Professor of Meteorology at the Massachusetts Institute of Technology, who believes (in remarks reported by *Science* magazine) that "a consensus view has been declared before the research has hardly begun". His opinion is that increasing levels of greenhouse gases will have only a small effect, mostly because of the moderating influence of water vapour in the atmosphere. In most models, raising the temperature of the Earth's atmosphere means that water evaporates from the oceans to increase the amount of water vapour. This exacerbates the greenhouse effect yet further because water vapour is itself a greenhouse gas. In Lindzen's view, the picture is more complex. Most of the water vapour which would go into the atmosphere would come from oceans in the tropics, and be driven upward in fast air currents. Here much of it would cool and return to the Earth as rain – and as the temperature rose and the air-current machine worked faster, the process of removing water vapour from the atmosphere would actually become more efficient.

Lindzen's views have been widely criticized by scientists who point out that the full range of processes acting to increase or reduce the greenhouse effect is far more complex than he seems to take into account. Whether global warming would dry or moisten the upper atmosphere is one of many questions which need answering, but the common-sense position that higher temperatures mean a wetter atmosphere is the one which most meteorologists seem to prefer. In more general terms, the argument is about mankind's ability to influence the overall properties of the atmosphere. In Lindzen's view there are so many moderating effects in the atmosphere that human action is unlikely to have much influence, while the mainstream view is that industrial activity on its present scale in the developed world is indeed able to affect the world's weather significantly.

Some of these issues will probably be resolved by experiments now under way, especially detailed satellite observations of the temperature of the Earth's solid and ocean surface and of the atmosphere, like the Earth Radiation Budget Experiment now being carried out by NASA. This would allow issues like those raised by Lindzen to be resolved. The models which these observations feed should also be able to sort out human-induced greenhouse effects from other short-term climatic changes which we now know too little about, like the effect of volcanoes on the atmosphere. Several major volcanic eruptions in the last century have been blamed for falls of up to half a degree in the temperature of the Earth's atmosphere, because the dust they throw into it reflects incoming solar energy which would normally reach the ground back into space. Since the hot decade of the 1980s was on average only about 0.3° warmer than earlier periods of the twentieth century, a difference of this size is significant in global climate terms. The massive eruption of Tambora in Indonesia in 1813 is blamed for the "year without a summer" of 1816, when snow fell in Europe in June and July.

However, the most significant task for greenhouse effect research is to determine just how greenhouse gases move in the Earth's atmosphere and just how much global warming will result from their increasing future concentration. For most of these gases, reasonably established science exists to tell us how long they will dwell in the atmosphere once released, and how much greenhouse effect they will cause. The main exception is the most important greenhouse gas, carbon dioxide, because it is a natural component of the atmosphere and has a number of important roles in chemistry and biology on land and in the oceans. Putting more of it into the atmosphere could have any of a range of effects – from increasing the rate at which limestone is laid down in shallow seas to allowing plants to grow more rapidly, since many types of plant are known to fare better with slightly more carbon dioxide to photosynthesize than is present naturally in the atmosphere. In practice, both effects will probably occur to some extent. The real requirement is to work out a complete carbon dioxide budget for the Earth as a whole. This would describe all the flows of carbon dioxide into and out of the atmosphere and allow a real forecast of future carbon dioxide concentrations to be built up from information on fossil fuel consumption and other sources of atmospheric carbon dioxide. Today's models do not, for example, account fully for the removal of carbon dioxide from the atmosphere by land plants, but may overstate the oceans' ability to absorb it.

Such a budget, when it exists, will allow a large number of policy options for dealing with the greenhouse effect to be tried out. They will also illuminate subtle weather problems which global warming might cause. These include the possibility (raised by Andrew Bakun of the US National Oceanic and Atmospheric Administration) that increased temperatures will mean an increase in foggy, cool weather along continental coastlines, by encouraging the upwelling of cool ocean water just offshore. This will be a nuisance for sunbathers but will also have a profound effect upon fishing, since such upwellings bring massive amounts of nutrients to the ocean surface and feed important fisheries.

New knowledge should allow us to discover whether the middle of the next century will be a degree or two warmer than the late 1980s – the optimists' view – or whether the warming will be several degrees, raising the Earth's temperature by perhaps as much as five degrees above those we are now used to. At the time of writing, some of the massive temperature rises of earlier predictions – those involving rises of five degrees or more by the middle of the twenty-first century – seem to be being scaled back in favour of figures in the 2–3°C range. As we have seen, there are plenty of unknowns in the science to push this figure up or down, but it is still worth thinking just what a temperature rise of this size might mean for sea level.

We have noted that there are several ways in which rising temperatures translate into rising seas. Of these, the most predictable is the overall increase in the volume of sea water which a warmer world would imply. (Because water vapour is far less dense than water, the increased amount of water vapour in the atmosphere would have only a minute effect on the rise in sea level caused by global warming.) After that, things get more complex. The first problem is the melting of what scientists casually term "small" glaciers – those glaciers other than the massive icecaps of Greenland and the Antarctic. Beyond that is the main event in sea level rise – the possibility that the massive icecaps of Greenland will melt and those of the Antarctic will themselves from the land and cause enormous sea level increases.

The basis for today's understanding of sea level rise is the observational fact – obtained from a wide range of tide gauge data – that sea level appears to have risen by about 10–15cm in the last century as global temperatures have risen by about 0.5°C. According to British climatologist Richard Warrick, a further temperature rise of perhaps 2.5°C would add almost 18cm to today's sea levels. Here, the sea level rise associated with the melting of small glaciers is likely to be minor, although the other effects of removing major areas of snow and ice cover from temperate and even tropical areas of the Earth will be large, affecting everything from drainage patterns to skiing holidays.

This looks messy enough, but now the uncertainties start to add up even more fiercely. As we have seen, it is by no means established that any sea level rise at all will result from global warming as it affects the Greenland ice areas. The net result might be a small fall as more water vapour is carried into the Arctic to fall as snow, or a small rise as snow and ice melt at a somewhat enhanced rate, or a delicate trade-off between the two. In practice, the best forecast is probably to write off any effect at all.

Much the same applies to the ice sheets of the Antarctic, with one reservation. If a temperature rise occurs large enough to unship the major ice shelves of the western Antarctic, nothing can prevent a massive rise in sea level well beyond the other effects we have looked at. There is no experimental means of working out whether a particular temperature rise will have this result. As well as this doomsday theory, it is also possible that global warming might increase the amount of meltwater at the base of these ice shelves and increase the rate at which they calve glaciers into the Southern Ocean, giving an increase in sea level of several centimetres – enough to cause alarm but not in the same league as the wholesale destruction of the Antarctic ice shelves.

However, this science contains a large number of unknowns. The most basic is our limited ability to measure the sea level at all. There are 1300 tide-measurement stations around the world, which at first sight ought to provide quite enough in formation about the level of the world's seas. Also of value is the fact that some of them have been producing data on a consis tent basis for at least a century, which makes it possible to draw conclusions about long-term trends in sea level. The problem is that other effects which have nothing to do with sea level rise can be large enough to mask any change in sea level. Tides can alter sea level by several metres in some parts of the world; seas onal temperature changes make sea water expand and contract; sea currents move water from place to place. Oceanographers think in terms of something called Mean Sea Level, but deter mining it is a serious problem. For example, the solid Earth itself is subject to tidal deformations of up to 50cm in a day, which means that sea level changes of tens of centimetres are invisible unless statistical techniques are used to separate them from the large distorting effects. Some time during the next decade this problem should be eased by new research tools, especially satellites carrying altimeters – in effect, very sensitive radar sets – able to measure the height of the sea below them every few kilometres to an accuracy of a few centimetres. Future satellites planned by the European Space Agency, NASA and others will gather this data at intervals of a few days across the Earth's oceans. As well as providing detailed general information, this will help many Third World countries with large numbers of outlying islands which do not have the scientific capacity to gather detailed sea level data from them all. In addition, a large network of tide gauges (the main one being called GLOSS, the Global Sea Level Observing System) and databases of sea level measurements are now being established by groups like the UK's Natural Environment Research Council to allow better information on sea level change to be obtained. The GLOSS system will produce data from areas like the Antarctic and the

Pacific, from which little sea level information is now received.

Also essential in assessing the future of world sea level is new knowledge about the way in which warmer weather might translate into higher seas. Part of the problem is that the global warming of coming decades is not going to occur in a uniform and fixed pattern across the Earth, any more than today's weather does.

Calculating just how climate change will work in practice will involve more work on just how the oceans and the atmosphere shift heat around the globe. A 1989 report from scientists at Princeton University in the USA casts the problem in an optimistic light, pointing out that temperatures rise in the southern hemisphere far more slowly than in the northern because of the immense mass of its oceans, which absorb heat, and because of the isolation of the Antarctic Ocean from the rest of the world's oceans at the so-called Antarctic Convergence. The result is that by the middle of the next century, increased atmospheric carbon dioxide could have made North America 3–4°C warmer while parts of the Antarctic are between 2° warmer and 4° cooler. If the major concern is the risk to the massive ice shelves of the Antarctic, this sounds promising. By contrast, a map produced in 1990 by the UK Meteorological Office and published in *Physics World* magazine shows the same areas becoming 8–10°C warmer, implying that there is still some way to go before our understanding of the issues is perfected.

In the same way, new methods of measuring the sea will allow us to find out significantly more about the way in which the seas might actually rise when and if they do. The Earth is a rotating, irregularly shaped lump of rock whose seas and oceans are confined in even more irregularly shaped basins and moved about by forces like heat, gravitation and wind. The seas are not actually flat because of tides and currents, and because the currents are largely driven by temperature differences they will change as the world warms up.

In recent years the use of radars on satellites has vastly increased our ability to measure the exact height of the sea. At the same time, oceanographers' ability to predict new current patterns means that they can say something sensible about how sea levels might alter as the world's oceans grow. Uwe Mikolajewicz of Hamburg says that even the simplest form of sea level rise – from thermal expansion of the oceans – is complex. He claims that a 19cm rise in sea level from thermal expansion in the next fifty years could mean that the Ross Sea in the Antarctic will fall while the North Atlantic will rise by 35cm – an unpleasant scenario by any standards.

In some areas, however, our understanding of the processes responsible for sea level rise is showing signs of improvement. For example, a 1990 report from researchers at the University of London's Mullard Space Science Laboratory suggested that observations from satellite-borne radar altimeters used to detect wave heights at sea are also showing up measurable thickening of ice in the Antarctic. This, paradoxically, is a bad sign, since the ice sheets are not completely stable. They stay in balance when the rate at which new ice is accumulating on them from snowfall matches the rate at which they lose material to sea in the form of icebergs. If a more humid atmosphere dumps more snow, it is even possible that the ice shelves will break off and melt, adding their material to the world's oceans and causing a massive sea level rise.

Similarly, new work by researchers at NASA shows that the Greenland icecap deepened by some 20cm a year between 1978 and 1985, and by another 28cm the year after. This seems to be consistent with the idea that the ice is growing as warmer, wetter weather feeds snowfall. But there are still major holes in our knowledge of the transport of water – as water vapour, snow, ice and liquid water – in the polar regions. For example, the rapid movement of Antarctic ice, which is lubricated by liquid water at the base of the ice, is only now being studied in detail. It is known that rapid growth in an area of ice cover can be a symptom of warming rather than cooling: it implies that water is being generated in larger amounts to lubricate ice movement. But the application of this knowledge, gained from small inland glaciers, to major ice sheets is still at an early stage. In addition, some potentially useful data, like some measurements of ice thickness made from space and many under the ice by submarines, are classified by governments because they are thought to reveal too much about the scope of equipment which might be used in anti-submarine warfare. A reversal of such decisions will be the proof that governments really do think that cold wars are becoming less vital to their interests than an overheating planet.

If they do, finding out more about the greenhouse effect and sea level rise will be only half of the task they and others have to face. The other essential will be to produce solutions to the problems the rising seas will cause. These answers are likely to come in three categories. Some sea level rise problems will be amenable to solution by clever engineering and other technological advances thought up to keep the sea out, or minimize the effect of its rise. Cities and large industrial sites are most likely to be protected by sea defences and other expensive capital projects, because of their financial value and political weight. Other solutions will involve a more thorough approach in which human activities near sea level are replanned as the sea rises. These differ markedly from the first group of solutions, those meant to allow something like normal activity to continue in rural areas of the world, which will probably be severely affected by sea level rise but tend to lack the money or political influence which capital-rich urban and industrial areas take for granted.

Even if it were possible to protect the whole of the world's coastlines from the rising sea with thousands of kilometres of flood defences, the problem would not be solved. In particular, such a drastic step would not solve the problem of salt water from higher seas finding its way into groundwater, rivers and the other water sources on which agricultural production depends. The increased salt attack on agricultural land will occur alongside other effects of global warming, which will shift vegetation zones polewards and allow crops (and weeds) to prosper in new areas while they become less successful in their existing habitats. Also important will be the effects on plant growth of the increased amounts of carbon dioxide and water vapour. Plants need both, and many, but not all, grow faster when the supply is increased – especially the supply of the former.

We have already met some plants which can cope with salt

water – the mangroves, a family of trees found in the tropics and capable of handling anything from brackish water to the salt concentrations found in the open sea. In addition there are marine plants – the most obvious is seaweed – which show that plants and salt are not necessarily incompatible. On the other hand, most of today's commercial crop plants cannot cope with more than a small amount of salt, which means that they are vulnerable to salt getting into the water supplies they feed on or being deposited on agricultural land by flooding, both likely outcomes of sea level rise.

In 1990 the US National Research Council published work showing that there are hundreds of plants capable of growing in salty or arid environments in may parts of the world. In the words of *Science* magazine, their investigators found "Israeli farmers irrigating specially bred tomatoes and cotton with saline water; Pakistanis growing kalar grass in waterlogged, high-saline soils as foodstuff for livestock; and Mexican farmers harvesting Salicornia, a succulent plant that thrives in sea water and produces a safflower-type oil". There is also the quandong, "an Australian tree that produces a cherry-type fruit".

It is probable that large sums of research funding will go into these plants in the next few years. It would be nice to report that the incentive will be to avoid impoverishing Third World farmers or to reduce the risk of famine. In fact, the main reason for spending the money will be to protect the incomes of farmers in the western states of the USA, whose land is being attacked by salt washed to the surface by irrigation. There are two possible ways in which knowledge of salt resistance might be used. The most obvious is to start growing salt-resistant plants in place of today's crops. However, modern technology allows more intriguing options to be researched. It is possible – at least in principle – to isolate the genes or groups of genes which bestow salt resistance on plants and put them into plants which at present lack such resistance. The US report points to asparagus, wheat, rice and barley as possible recipients. In addition, it has proved feasible to use more traditional breeding methods to introduce salt resistance. One group from Weber State College in Utah crossed a seashore-dwelling tomato from the Galapagos Islands in the Pacific (which tasted pretty terrible) with other breeds to produce a tastier type which still prospers in sea water.

But the possible damage to agricultural production from sea level rise is only part of the story. Rising seas also threaten natural ecosystems and the plants and animals which constitute them. Such systems, despite the air of timelessness they display to human visitors, are constantly adapting to new pressures, and a tactic of attempting to preserve them in an unchanging form is not likely to succeed. The most notable characteristics of natural systems are their complexity and their diversity, which means that it is impossible to generalize about their preservation except to say that each needs to be investigated on its merits. However, the kind of areas which are likely to be threatened by sea level rise may well share characteristics which make them especially worth preserving. Some have useful roles in protecting inland areas from storms and flooding – like the mangroves and coral reefs we discussed earlier. Some have direct economic value – with care, mangrove trees can be cropped for fuel, and they shelter a wide range of edible animals and plants. And, like the threatened rain forests of South America, they also contain unknown treasures of plant and animal species which might be of particular value in coming years – if only for their ability to resist salt attack.

However, the biggest sums of money which are likely to be attracted to the problems of sea level rise, and the most focused political attention, will not be spent in mangrove swamps or tomato fields but in cities. The possible sea attack on places where capital investment, human habitation and history, and commercial and political influence are concentrated will involve the most delicate choices about means of dealing with the rising sea.

The real question – and the one calling for the most careful research into the options – is the extent to which it is possible to defend against rising sea levels. Some experts suggest that there are many cases in which a policy of retreat is the only possible response, especially if seas rise to anything like the extent anticipated by some especially pessimistic forecasters. Many coastal cities already spend large amounts of money on sea defences. The sums which West Germany spends on keeping the North Sea out of Hamburg would be beyond the scope of most Third World nations, but is regarded by Germans as worthwhile because of the massive wealth (and tax base) of Hamburg and because the alternative – abandoning a historic and thriving city – cannot be contemplated.

On a world scale, the same approach cannot be adopted universally. Instead, we shall probably see a gradual process of change whereby more and more attention is paid to salt incursion, flood risks and other effects of sea level rise when investment decisions are considered. The first decisions to be affected by such logic will probably be concerned with industrial plant rather than cities. At the moment, rising sea levels do not rank with tax breaks, the availability of skilled labour and other more conventional factors when decisions on where to put new factories are taken. In future, the possible effects of salt intrusion on water supplies, or of flooding, will enter such calculations more and more. The same considerations will also form an increasing part of decisions on siting new urban development. In the longer term, it is also inevitable that the same thinking will pervade decisions about redeveloping existing cities – although it would be a tough planning department which advocated abandoning a millennium-old capital city in favour of a new site uphill merely because rising sea levels were making life there tough for the poorer inhabitants.

This means that the main mechanism for dealing with sea level rise in most locations is going to be engineering rather than ingenuity. In many cases a strictly technical assessment of the options might suggest that such a strategy is bound to fail in the long term. Such spending has the political attraction for local and national governments that coastal defence works are a visible sign of a commitment to an area and its population, and that they provide immediate defence against normal flooding as well as against the longer-term effects of rising sea levels. Even today, floods and surges are estimated to cost as much worldwide as earthquakes, and are one of the major world hazards to life.

The current debate about just how to protect low-lying land from the sea centres around a disagreement analogous to the contrast between different schools of thought on Third World development. In the latter case the argument is about "sustainable" development, using local skills and resources to provide for the needs of local communities, and the now less fashionable route of using external capital and high technology in an attempt to force areas and communities to turn into a version of Western society. Third World development experts may be cringing at this oversimplification, but the comparison with the state of the art in sea defence is a close one.

In this case, the failings of the old way of doing things are pointed up by the catastrophic flood of 1990 in the small town of Towyn in Great Britain, caused by the collapse of a sea wall under attack from a heavy storm. The wall was 8m high and over 2km long, and had been built by a railway company 140 years earlier. It had been put there to confront the sea, and because of the competence of its engineering had done so successfully for over a century. But when it failed, it did so completely and without warning.

Defences like this are in place around over 40 per cent of the coast of England and over 20 per cent of the coast of Wales, where Towyn is situated. Putting them up involved massive investment, and once in place they require constant maintenance to keep them working. In the UK and other developed countries, technologists are now starting to accept that they may not be the right way to set about protecting people from the sea. In Third World countries, governments do not have the capital for such an option to be contemplated.

One problem with existing major sea walls is that they have all been built to cope with existing sea problems. They do not have built-in slack to allow for sea level rise. They also need to be maintained and repaired. But their real disadvantage, according to sea defence thinking now emerging, is their all-or-nothing way of working. When they give way, they do so completely. And they are bound to fail sooner or later because the task they are meant to carry out – absorbing the energy of waves and currents – means that their physical structures are subject to loads which in the end will wear them out.

Some high-value locations in countries around the world will probably always be defended from the sea by expensive capital works. As we have seen, an example is London's Thames Barrier, without which the city would by now be at regular risk of flooding. But when the Thames was plugged by the Barrier, a considerable part of the total cost was spent not on the Barrier itself but on flood works downstream of it—otherwise the water which was prevented from going upstream would simply have flooded land somewhere else. This points up the other problem of solid flood defences: they may absorb the energy of the flood but they cannot absorb the water itself. Instead, they simply send it elsewhere. This means that it is very difficult to protect large areas of low-lying land by means of massive civil engineering works. Instead, attention is now being paid to technologies which imitate or assist natural flood protection.

Build a wall on part of a coastline which has beaches, dunes or other natural protection, and the effect will probably be to increase the rate at which diverted flood water eats away at such natural barriers. More sympathetic forms of flood defence depend on building a complete understanding of a flood hazard before attempts are made to shut it out. It is necessary to develop a computer model of a coastline and the seas around it, including tides and currents, under both normal and flood conditions. Groups in the UK and the USA are active in this research, in which three-dimensional models of the sea are built up showing currents, water depths and wave heights over time. A preliminary example of this approach in action is an experiment carried out in England at a beach called Jaywick Sands in Suffolk. The beach has always had a wall running out to sea to prevent longshore drift, the process by which sand that would otherwise guard against erosion is swept off the beach by tides and currents, but this wall had ceased to be effective. At Jaywick the problem was solved by building an art-ificial sandbar offshore to catch the sand removed from the beach as the tide went out: the next incoming tide washes it back free of charge. Another British experiment, this time at Pendine Sands in Wales, has involved planting fences on the beach to trap wind-blown sand, again building up the beach by reducing the rate at which it would otherwise be eroded away.

On a larger scale, US experts on sea defence have pointed out that coastal wetlands, which can be endangered by rising sea levels, are so called because they are wet. Provided that they have some protection from massive storm attack, perhaps from sand dunes or offshore islands capable of containing the full energy of the sea, they can absorb flood water and simply let it run back to the sea as the water level eventually drops. The same applies to the Everglades mangrove swamps of Florida, regarded by Floridans as an essential part of the state's sea defences. Among the economic activities they shield is the growing of over a million tonnes of cane sugar a year. However, the sugar-growers also use hefty quantities of fertilizers, which have been accused of overloading surface waters with nutrients that encourage the growth of algae and drive out normal mangrove species. In 1990 the *Financial Times* reported that a new research project on cleaning up the sugar industry's emissions might be a solution to the problem, although the industry is wary about the costs and is making the noises customarily expected of an industry under attack from environmentalists: the case is not proven, and conceding the point will simply stimulate more demands from the enemies of sugar production and other forms of industrialized agriculture.

However, the gentle approach to containing sea level rise is a long way from becoming established wisdom. In the USA, according to the International Programme on Climate Change, spending a mere $160 million a year on conventional sea defences would provide protection against a 1m sea level rise and yield over $2 billion savings a year. By contrast, the equivalent figure for Bangladesh is over $300 million – for financial savings a mere fraction of the size, since US land is so much more valuable and productive than Bangladeshi. So the big research money is bound to be spent on means of protecting high-value land in the developed world. This may be why the Bangladeshi flood protection Action Plan looks remarkably like a programme thought up for use in the developed world and transported wholesale to Asia. It involves building hundreds of kilometres of flood defence walls, moving hundreds of thousands of poor people who have nowhere to go, and putting in massive drains to shift cubic kilometres of flood water from river to river. No sign here of alternative approaches like assisting natural circulation of flood water and natural defences against it, much less of ideas about reducing the loss of tree cover in the Himalayas to cut down the amount of sediment coming downstream to block the Bangladeshi river system.

Perhaps a more encouraging approach to the problem is the Cities on Water process started at a conference in Venice in 1989. The aim of the first meeting was to encourage cities and regions with a common interest in the sea level rise problem to talk to each other about the issue, with one eye on the fact that rising sea levels which threaten cities, especially in the Third World, are aggravating already intense problems of population pressure and lack of basic services and resources. The organizer of the event, Professor Roberto Frassetto, says that there should be a ten-to-fifteen-year programme of work designed to develop human, engineering and environmental approaches to the problem, build links with other forms of research on global climate change, and gather data; and a further period of perhaps five years for key decisions on matters like defending or abandoning areas and removing the uncertainties from the sea level rise predictions. After that there would need to be twenty years of implementing solutions. To start with, cities with especially severe problems would be used as prototypes for the solutions which were later adopted elsewhere.

Apart from Venice itself, candidates represented at the conference include Osaka in Japan, a city built on water which has already spent very heavily on flood defences in recent decades, and African cities like Lagos in Nigeria, Banjul in Gambia and Dar-es-Salaam in Tanzania. Each of these has already been subject to sea attack on harbours, tourist areas and business centres. According to A Chidi Ibe of the Nigerian Institute for Oceanography and Marine Research, the need is for cheap measures to tackle flooding and erosion, which now exist mainly as ideas rather than usable technologies.

Whatever the future for sea level, the future for ocean and climate researchers, developers of sea-proof crops and inventive sea defence technologists may seem from all this to be assured.

But their ability to produce good work will be valuable only if they have funds enough to carry out first-class work well, and spending the money will be of any use only if the policies adopted to deal with sea level rise make the most of the knowledge such research produces. In the last chapter of this book, we shall look at the political issues which rising sea levels bring with them, and at the possibility that the problem will be solved by political action in the developed and Third worlds.

Guyana

Guyana is a large and sparsely populated country, with 900,000 people in an area about the size of Great Britain: 215,000 square kilometres, but most of them live in the country's low-lying coastal plains. According to work carried out for the Commonwealth Secretariat by R.F. Camacho, just 3 per cent of the country's area houses over 90 per cent of the population and is the site of over 70 per cent of GDP. This is regarded as the only area of the country with significant potential for agricultural production. The only major industry outside the coastal strip is mining in hilly areas inland.

The coastal area is up to 16km wide and is one of the largest expanses of artificial land in the world. It was created by the energy of slaves and the hydraulic engineering experience of Dutch settlers, in the eighteenth and nineteenth centuries. The original plan was to produce new dry land from swampland and the sea, a massive effort economically justified by the profits from sugar grown on the new cultivable land. Over the years the original inland polders have become the coastline as the sea has removed the beach in front of them, and the sea coast is now protected by geotextiles and boulders. There is a Sea Defence Board with wide powers.

A rise in mean sea level of, say, 0.5m would mean appreciable economic damage for the Guyanese economy, and a rise of 1.5m would call for forced drainage at massive cost in both the growing areas and the inhabited part of the coastal strip. While rapid price changes and fluctuations in exchange rates make calculations difficult, it seems that the economic activity in the coastal strip is worth some $800 million a year.

Tactics for dealing with the threat to this area are difficult to devise. A policy of shifting the Guyanese economy inland is impractical; there is only one higher area suitable for agricultural production, and even this is better suited to ranching than to crop cultivation. It is just possible that a new Guyanese economy could be created inland if successful searches for oil led to industrial and energy development there, but this hypothetical future cannot be relied upon and policies are needed instead to protect the coastal zone.

This will mean a programme of new coastal defence works along virtually the whole of the Guyanese coast, starting in areas which are already coming under attack and continuing for thirty to forty years. Some coasts now protected by natural reefs offshore will need artificial protection as well. In addition, the land's natural and existing artificial drainage will need to be expanded by the addition of new drainage and pumping capacity. This programme would cost about $22 million in its first five years of operation, when technical studies and preparation would be the main element, with more to come later according to the precise amount of sea level rise envisaged in the following decades. Much of this money might well have to come from development aid, since the costs would be incurred in hard currency.

These defensive works would also involve a continuing cost for the Guyanese economy even if sea level rise were stopped at some future stage. Areas where cultivation was able to continue only because of the use of artificial pumped drainage would have to go on using fossil fuel, and the new sea defences would require continued maintenance. The building programme for the first five years – the only part planned in detail by Commonwealth consultants – would involve creating about 4km of new sea walls a year and 1.5km of restored walls. This massive effort would also require a new project management organization within the Guyanese government, and a large research effort to find out more about the scope of the problem and design and maintain new sea defence structures.

7 Living with the rising seas

For the oceanographer, the solar physicist, the atmospheric chemist or the glaciologist, sea level rise is a subtle sideshow to the greenhouse effect. Ask an expert in sea defence about it, and the problem will probably be described in terms of keeping water and land in their present positions. Ask planners, agriculturalists or applied scientists, and it is a new development calling for practical solutions in farming, land use, urban design or some other area.

Talk to a conservative politician and the issue involves balancing the unknown costs of environmental damage against solutions which are bound to be expensive. Try a more radical one, and it is yet another case of the habits of the rich nations causing suffering to those who can least afford it in the Third World. Ask someone without a specialist interest in the problem what they think about sea level rise, and you may well get little more than bafflement in response.

All these answers are correct from the point of view of the people giving them. The sheer size of the issue means that many different views point to policies – including public information and education – which might usefully be followed if the worst effects are to be avoided.

However, it is certain that the issue will in practice be tackled in a less scattergun fashion. The places where sea level rise is already being taken seriously are the low-lying island states of the Third World. These states will probably continue to be the countries which put sea level rise at the top of their domestic political priorities. This is doubtless the correct response. For such states, continued existence itself is in the balance. Since most such countries are already poor and have few national resources, whether the asset in question is mineral wealth or skilled engineers, coping with sea level rise on top of their other problems is like being asked to play a piano concerto while fighting off an unfed tiger. It is certainly a responsibility of the richer nations to make sure that such states' calls for money and technical assistance are met in full.

The practical task for low-lying countries is to reduce the effect of rising sea level by reducing the sea's attacks upon land, buildings and water supplies. This can be done by conserving natural sea defences like beaches, mangroves and coral reefs, and by preserving water resources by water conservation programmes. In addition, even low-lying states are not usually in danger of complete disappearance below the waves, but they do tend to have far too many of their major assets, from airports to power stations, at very low elevations above sea level. Some of these may have to be shifted to keep them from extensive flood damage. In addition, the main cities of most such countries have been growing because of migration, putting more stress on confined areas which lack the machinery to cope with it. Migration of this type is endemic to the Third World but might be slowed in low-lying states by proof of national and international action to combat the sea level rise problem. At the moment, anyone already on the economic margin is going to be encouraged to move by proof that sea level rise is finally pushing their way of life over the edge. The same applies to Third World countries threatened by significant land loss as the sea rises. Examples we have looked at range from Bangladesh to Guyana, and the problems are as diverse as the locations.

However, a common factor in all cases is the extreme pressure on such countries' slender reserves of economically viable land. Making the most of it by reducing migration is bound to be a part of a systematic policy of dealing with the sea level rise problem. Some countries, like Guyana, can also respond to the pressure by spending money on sea defences, although it would be hard to find any easy answers in the case of the immense problems faced by Bangladesh.

In other Third World countries, however, it is possible, at least in principle, to develop policies of shifting highly valuable

Figure 7.1 **Areas vulnerable to rising sea level**

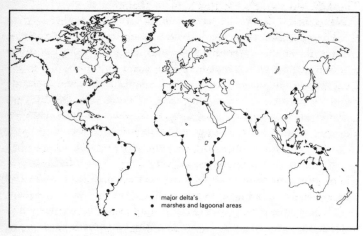

Source: Environmental Protection Agency

assets uphill, moving commercial and bureaucratic centres and generating employment away from today's sea level. In addition, the shift of people to towns – especially to capital cities, which are usually coastal – can be attacked by land and farming reforms which should be begun even if the sea level problem did not exist.

In the developed world, the problems arise in a different way. The examples of Hamburg, Venice, the Mississippi delta and London show in different ways that developed countries are willing to devote energy and cash to shutting out the sea, on a scale which would not be available to the Third World. Even more extraordinary is the case of the Netherlands, which has not only protected the bulk of its land area from the sea over a period of centuries but added to it by reclamation. In a country which exists by virtue of control of the sea, the choice of fighting or retreating arises in a particularly stark form. The Dutch have the skill and the wealth to support a decision to fight, unlike many Third World countries with a similar dilemma.

Nevertheless, even rich countries cannot escape the need to think hard about what sea level rise means to them. The issue really falls at the junction of planning and politics, because it involves a series of judgements about what might happen to sea level, how it will affect low-lying areas and just how the possible decisions about defences against the worst effects might stack up against other spending priorities and the interest groups which support them. This planning is likely in practice to form part of wider exercises in which the whole of a national response to global warming is thought through. In urban areas, sea level rise will be the main problem in many cases. It is mainly cities, towns and industries which can, at least in principle, be defended from the sea in a manner which allows the investment to pay for itself.

However, a sufficiently subtle planning exercise will take into account just what will really be acheived by defending cities against sea level rise, not merely the problems which might arise if spending on the activity is restricted. One place in the world which is endangered by rising sea levels, where protection from the sea would attract all but universal support, is Venice. But as we saw, the act of protecting Venice from the sea involves an inherent decision to send the flood water somewhere else, amplifying the sea level problem for hundreds of kilometres along the Adriatic coast.

This kind of analysis of the full costs of sea defences indicates that the Greater London Council might have got it right in the 1960s when the Thames Barrier was built. There, a significant part of the cost was incurred on coast defences and river protection downstream from the Barrier itself. As we have seen, this means that when the Barrier is in use, the water which is shut out of London does not spill over and flood somewhere else instead.

London is rare among sea-level cities in having only a small river, connecting it to the sea via a lengthy estuary. This means that with enough money, a single barrier can be used as a protection against sudden inundation. The problem arises in a more intractable form for coastal cities with a more conventional geography, where there is a lengthy waterfront which cannot be shut off from the sea. Here the problem is to decide how the sophisticated machinery of a modern city – capital equipment, people and social structures – can stand up to the threat.

Viewed in this light, the prognosis seems to be poor. The whole point of a city is that it is a single social unit. In practical terms, this means that people move around it from district to district, power supplies and transport are linked together across it, and its physical support mechanisms like water and sewage are single, complex systems which cannot be divided into separately operating units. The same goes for human activities like living, working and fun. None can be isolated from the others in a single city – although to add to the sea level rise problem, most cities have grown from small beginnings at the water's edge, so that their most exclusive living areas and their most important commercial activities take place nearest the sea. This means that it is not possible in most cases to have a policy of selective abandonment for cities. A decision that the low-lying areas of a town are expendable is inherently a decision that the whole city is no longer viable. The area in question would turn into a sea-level ghetto of poor housing and businesses, with anyone who could afford to be somewhere else getting out. Services running through the area would be cut and pressures on social structures increased within the affected zone and elsewhere.

The alternative, for cities which cannot isolate themselves from the sea, is to develop policies which allow the city to defend its most valuable assets, protect its population and keep its social structures intact. The problem is bedevilled by the fact that most big cities are magnets for migration, internationally and within states. This means that few have a plentiful supply of nearby land for expansion uphill as the sea rises. In addition, as the participants in the 1989 Cities on Water conference noted, the cities in question are already under pressure from the sea, with salt invading buildings and water supplies, storms hitting them with increasing frequency, and seafront areas under attack from water pollution.

It is difficult to imagine solving these problems through conventional political processes. After all, the worst effects of sea level rise are still several national and local elections away from being felt with their full force. In the meantime, as Cities on Water participants pointed out, the switch away from the heavy engineering approach to sea defence in favour of soft solutions is certain to mean a change to ideas which do not create votes by producing big-money engineering projects or visible, spectacular results like massive stone walls placed between the electorate and the threatening sea. By contrast, the mayors of cities which are planning for the future will have to make themselves unpopular, for example by resiting key roads and railways uphill from their present position, sending industries which demand huge amounts of water elsewhere when they are looking for sites, and putting cash into apparently mundane investments like sewage works and water supplies sited above the higher sea levels of the future.

Only a serious planning effort and clever political thinking will allow programmes of this kind to be carried through, but there is an up side. Once a city becomes known as a potential sea level rise victim, with disrupted public services, flood-prone areas, water supply and sewage problems, and rising land values caused by pressure on habitable areas, it will cease to be attractive to investors, skilled labour and success-minded governments. Local and national politicians who have any sort of eye on their own careers – or even a streak of public conscience – are bound in future years to regard avoiding this kind of problem as a political priority.

Move away from urban areas, and the problems arise in a new form. The main differences are the lower value of rural land, its lower population density, and the fact that rural areas have a single economic activity – farming – on which everything else depends.

Defending the agricultural activities of rural areas is the key to dealing with the sea level rise problem there. This is the part of the sea level issue where the interests of people in the First and Third worlds are closest together. In many parts of the Third World, land damage is already acute, land ownership is highly unequal, migration to towns is endemic, and there is a general feeling that cities are the main focus of political interest and public spending. In the same way, farmers in the developed world feel – especially with reduced spending on the European Community's Common Agricultural Programme – that their interests are rarely put at centre stage by ambitious politicians.

Adding sea level rise to the problems which farming already faces worldwide is bound to accentuate all these difficulties. It will naturally put some land out of use, adding to the incentives for rural people to move to towns and thereby causing problems in both city and country. But as we have seen, the real problem is not, in most cases, that land will vanish underwater entirely. Instead, the risk is of flooding, salt pollution and soil damage which turn gradually – if they occur – into complete destruction. This pattern has an up and a down side.

The up side is that in most threatened areas, there is time to think about policies before the problem becomes critical. The need is for a version of the Cities on Water process (or what that process will turn into if it is successful) whereby thinking is done before the problems becomes acute.

In the case of agriculture, the main problem to be solved is the creeping reduction in food output as the sea attacks low-lying land. For example, nobody yet knows just how large the promise of the salt-tolerant plant species mentioned in the last chapter really is. It would be worth a considerable outlay to find out whether credible saltproof versions of the world's major food crops can be produced, and if so, what the drawbacks of using them are likely to be. It could be, for example, that they can be grown only with unacceptably poor yield, or with the use of massive amounts of artificial fertilizers. As well as being environmentally undesirable, and all too often harmful to the eventual consumer of the food, the latter are simply beyond the financial means of all but the most prosperous Third World farmers.

It is worth the effort of answering these questions partly because the salt pollution problem has ramifications far beyond islands and coastlines threatened by the rising seas. Large parts of the developed world – most famously California – also have salt pollution problems, mainly caused by irrigation water flushing upwards salt which would normally lie at depth in the soil. This is also said to be the problem which led to the decline and fall of the Mesopotamian Empire in the fourth century BC, and there are still extensive areas of the poor world, notably in Asia, where it is endemic.

As well as looking at the salt problem from the point of view of plants, it will be necessary to think about it in terms of animal husbandry: like people, farm animals need salt, but only in small amounts. As with plants, there are plenty of animals which tolerate salt, including mammals like the whales. Amongst the more gung-ho biotechnologists are those who claim that the genes for salt tolerance can simply be shifted from mangroves and other salt-tolerant plants to normal species: there is no reason to think that the same arguments could not be applied to animals. However, the cynics who say that the whole matter is more complex than that are probably right, at least at this stage. Salt is one of the commonest chemicals on earth, and living things have been evolving in its presence for hundreds of millions of years. The main reasons for avoiding it – the effects of the pressure of dissolved salt on cell walls – are inherent to the basic structures of plants and animals. In the same way, salt tolerance is unlikely to be transferable from species to species by a simple shift of one or two genes between the two – even if shifting a gene ever becomes simple.

At the same time, there must be more work on the economics of rural areas under pressure from rising sea levels, and a new assessment of their value to the countries of which they form part. As well as the monetary value of the things they produce, including food, this would involve looking at the monetary and moral value of their wild areas and at the possible future value of resources like unknown species – and potentially useful genes. One possible result could be to conclude that not all areas threatened by the sea are precious enough to be protected from sea level rise – which, after all, is itself a natural process which human activity is speeding up, not a completely new side-effect of twentieth-century industrialization. This calculation will naturally be complete only if it also contains an accurate account of the costs of defending rural areas against the sea. This cost is itself inherently hard to calculate. The new tendency towards subtler methods of sea defence engineering may lower the costs, but only if money is spent to improve the methods involved and make them plausible to the people they are meant to protect.

The scope for protecting low-lying land and the agricultural activity which occurs on it from sea level rise is broadened by the slow pace at which the rise occurs, but the down side of the equation is that the rise occurs too slowly for most political processes. If a big and complex problem is more than one or two elections away – cynics say – there is little chance of its impressing itself upon politicians as a spending priority. Even more highly qualified cynics would add that in many of the countries threatened by sea level rise, the politicians have arranged matters so that even elections do not trouble them.

However, there are ways for people affected by sea level rise to make a case for quick action. As we have seen, if sea level rises by 50cm in the next half-century, there are few places in the world which will suddenly turn from land to sea. It is possible for regions or interest groups to argue the case for short-term spending to relieve the pressures of sea level rise, like shifting water supplies above the danger zone, in terms of their immediate value as well as their long-term role against this rise. However, it would be wrong for countries to start taking even this reasoned approach to the issues without thinking hard about the long-term problems they are making for themselves by doing so. No amount of spending on countermeasures can stop the seas rising – all it can do is mitigate its worst effects. If the sea goes on rising and the people confronting it are able to point to a past history of public spending to keep it out, the case will be made, by normal political standards, for an indefinite programme of spending even as the value of the land under attack and the food it produces falls and its ability to support human communities is diminished.

The alternative to this approach is the most difficult route for communities and politicians to contemplate: to think through in detail a policy which may involve abandoning land altogether in the face of rising sea levels.

The possibility of admitting that a part of one's country, or a piece of land which one has worked hard to make productive, may at some point become untenable because of an apparently arcane geophysical effect to do with air pollution is most unlikely to commend itself at first sight to anyone. But it is possible to sketch out a basis for establishing cases in which the best policy might indeed be to admit just that. The easiest case is that of land which is genuinely not cultivated, even informally, where the pressure for protection is likely to be at its weakest. For such areas, the correct decision might be that unless genuinely irreplaceable natural assets, such as a unique species, are in danger, the sea will be allowed to rise without impediment. In most wilderness areas, this decision would be made easier by the fact that the sea would not threaten to inundate more than a small part. Also relevant to the policy decision is the fact that shutting out the sea would cost money, which could be spent on something else.

Beyond this policy of abandonment by benign neglect comes a whole series of possible decisions to relinquish to the sea land which someone owns and which may be economically productive. One basis for taking a decision to abandon it is the traditional accountant's one of lining up its productive value on one side of a piece of paper and the cost of protecting it on the other. In a world of low commodity prices and agricultural surpluses, this calculus might of itself suffice to rationalize the decision.

However, there is a more subtle set of reasons for supporting a policy of land abandonment in a limited number of highly specific cases. Most land areas do not abut the sea directly – except when they rise very steeply from it, in which case the problem is not so serious. The usual pattern is for there to be a beach, some wetlands, a set of dunes or some other barrier between the two. While this zone has little economic value (outside resort areas) it is the key intermediary between the sea and the land. Rising sea levels will increase the pace of erosion in this zone without necessarily doing anything to increase the deposition of sediment there. This means that preserving this vital zone will involve setting aside areas for it to expand upward and into, or facing extensive flood damage as the sea attacks former inland areas directly.

This means that new thinking about environmental economics, especially the value of land which produces no saleable output, is of special relevance to policies which might involve abandoning land in the face of sea level rise. Part of the problem can be solved by asking the public to attach an informal value to unproductive but important areas like wetlands, perhaps in terms of their role as breeding grounds for birds. Even this kind of calculation is complex. It would value a comparatively unproductive area of Scandinavia far more highly than a unique site in Africa because of the difference in affluence between the two regions. Even this, however, is an exact science in comparison to the process of deciding to abandon land which does have a provable economic value. In any case the alternative of committing nations to fight a steadily more unwinnable war against the waves should not become the accepted policy simply by default.

A decision to abandon an area in the face of the rising sea would be a difficult one even in an otherwise ideal world. In practice, it would be especially hard because the area abandoned would inevitably be one which was already under economic and social pressure anyway. In the Third World the pressures are at their most visible, with poverty, disease and migration to make them apparent. In the developed world the symptoms are less unmistakable, but it is still likely to be areas which are socially and economically at the margin which are pushed off the edge by sea level rise.

This means that decisions to fight rising sea levels – or let them take their course – will both call for the social structures of the areas involved to be strengthened. In some areas of the world, this could be less painful than might be imagined. In the European Community, for example, the funds spent by the European Regional Development Fund and under the EC's structural programmes might be used to pay either for defences against the rising seas or for the costs of moving people, industries and infrastructure out of their way. At the moment, the main focus of ERDF spending is the building of regional resources like trunk roads, airport runways or telecommunications links.

The idea is that industrial development will build up, emigration will slow and local economies will prosper once it becomes apparent that real money is being committed to areas which have previously been economically marginal.

But it is a considerable jump from this sort of spending to a decision to use money to pre-empt or solve sea level rise problems with a gestation time of decades. The only rationale for doing so would be the prevention of far more expensive problems later. When spending agencies consider the sea level rise issue, they are most likely to decide to respond either by pushing money into defences against it or withdrawing money from areas endangered by it. These would both be misguided priorities, since defences are not guaranteed to work and removing resources from affected areas is certain to make the position of the communities in the front line even more untenable than it would be anyway.

Instead, the case of sea level rise is one where the classic bureaucratic trick of carrying out more research instead of actually doing something might, for once, be the right choice. For this purpose, the task should be to find out just what the importance of sea-threatened areas is, in terms of their populations, their economies and their cultures, and their natural and ecological assets. Gathering this information is a quite different task from the parallel scientific one of assessing where sea level rise will probably be most damaging and where, how and at what cost it might be fought. The idea is to assemble a real assessment of the costs of reinforcing or abandoning particular threatened areas instead of taking, in an information vacuum, decisions which cannot be untaken later.

The same kind of logic should prevail, at least in translation, in other parts of the world. The problem will probably be seen at its starkest in the Third World, where even more distinctive approaches are called for. One thing which the rich world has not hesitated to supply to its poor neighbours is a flow of development experts to produce learned reports on the in problems. Consultants' reports on sea level in the Maldives probably exceed in weight the available information on the same problem in the USA, and in years to come sea level rise is bound to impress itself on local and national agencies in Third

World countries as an issue in national and local choices about the future.

In the Third World there are, practically by definition, few places with the kind of loose money available for capital investment which would be needed for complete programmes of sea defence against rising sea levels. However, a thorough look at the costs and benefits might reveal that in many cases, the defensive works involved could be used as one tool in the fight against unemployment and underemployment. So far, experiments like planting fences to encourage sand dune formation have been carried out mainly in the UK, but in future they might be seen to be relevant to the technical requirements of Third World countries, as well as to their ample supplies of labour in need of new skills and jobs.

The problem, however, is not one which can be solved unless the social stresses which sea level rise will mean for the communities it affects are made apparent in assessments of the problem. In the developed world, these stresses tend to manifest themselves in the shape of falling land values, increasing insurance costs and dwindling local communities – for example, in areas like the parts of Lincolnshire in England affected by serious erosion in recent decades. Areas like these feel that they are already under pressure and are likely to become less of a priority rather than more if the feeling grows that they are doomed by rising sea levels to suffer steadily more seriously from floods and erosion. In Third World countries like Bangladesh or the islands of the Pacific, or in threatened Third World areas like the deltas of the Nile or the great rivers of Asia, the position is analogous but the sheer scale of the problem is even more forbidding.

So in the end, the issue is one of equity. In so far as the problem of sea level rise is a subset of the more general problem of global warming, it is one which the rich world is imposing on the poor world. The exotic greenhouse gases like CFCs are almost entirely the product of activities in the developed world. And although countries the world over use oil, coal and gas, the sheer volume used in the developed world means that the carbon dioxide they produce is mostly emitted from chimneys and vehicle exhausts in Europe and North America. The only exception to the general rule that greenhouse emissions come from the developed world are the massive emissions of carbon dioxide resulting from the burning of tropical forests. Here the complications are immense, but it is certainly possible to argue that without the need to produce cash crops to service debts owed to banks in the developed world, the pressure on such lands would be usefully reduced. The reasons why the tropical rainforest ought to be preserved run far beyond the greenhouse effect, and include the species they contain, their human cultures and their uniqueness *per se*, but the argument in their favour most likely to appeal to Western policymakers is certain to be their role in removing carbon dioxide from the atmosphere and the immense amounts of it which are produced in the course of their destruction.

From this point of view, the problem is one of applying the "polluter pays" principle, which is enshrined in many countries' national legislation on environmental damage, to a problem of planetary dimensions and decade-long timescales. Rudimentary political principles imply that the first essential step is for the people most severely affected to get angry and make a noise about the problem. So far there are too few signs of this happening, although hopeful straws in the wind include the Commonwealth nations' concern about the issue, the interest which development groups in the Pacific have taken in the matter, and the Cities on Water lobby on behalf of regions and cities threatened by sea level rise.

In future it is essential for developed world political parties, aid agencies, environmental groups and individuals to place sea level rise among their central concerns. This would have seemed a quite arcane request just a few years ago, but now seems almost mainstream. And the speed with which issues like the hole in the ozone layer, which springs from a far more complex set of scientific problems than sea level rise, have become politically important implies that the same can happen with other environmental challenges. In any case, sea level rise is part of the greenhouse effect, which is already in the political arena.

Action on sea level rise is also easier to achieve because it is a global problem which affects particular parts of the world in a highly local fashion. The greenhouse effect itself is a subtle matter of parts of the world warming at a variable rate with variable effects. Any estimate of its actual costs – and even more so of its non-economic effects – would produce massive numbers with almost equally massive margins of error. But it is possible to tell with some precision whether a particular island, river delta or coast is under attack from rising sea levels – indeed, millions of dollars are being put into equipment to provide hard data on just this subject. At the same time, the subtler effects of sea level rise, like attacks on water supplies, are also visible to those affected. The only real grounds for disagreement are over whether a particular catastrophic event, like a major flood, is within the usual range of weather or ought to be attributed to the greenhouse effect.

Since there is no way, even in principle, of answering this question, the correct approach is to regard such cases, especially where they cause major damage to poor communities, as sea level rise precursors deserving not only relief but, more importantly, serious thought about their implications for the communities involved. Because of poverty and the demand for land, people in the Third World already live in places, like steep or flood-prone areas, which would be regarded as dangerously unsafe in the developed world. The threat of sea level rise is simply an extension of existing dangers for many Third World communities. In the long run, the most positive outcome would be for the problem to lead to a more complete appreciation of the hazards of life in the Third World and the new patterns of development needed to deal with them. This applies with special force in the case of rising sea levels, where the problem is bound up with other Third World issues like erosion, soil loss, forest destruction and unequal land ownership. Adding a distinctly new peril, caused by the actions of rich people in the developed world, to existing ones ought to be the occasion for serious thought about the existing dangers of the Third World as well as the novel ones.

The correct approach to the sea level rise problem will probably involve combining case-by-case action in endangered areas with an attack on the roots of the problem, the greenhouse effect itself. The greenhouse effect is such a colossal issue that action taken to counteract it will not arise solely from concern about sea level rise. This could mean that the sea level becomes a sideshow to other policy problems, but it has the positive aspect that the chances of real action being taken are increased by the other benefits which would accrue from a worldwide attack on the greenhouse effect.

A further problem is that climate change and sea level rise are only partly caused by human actions. We have little say in most of the ways in which the Earth's climate changes. The sinking and rising of parts of the Earth's crust, or the variation in the Sun's energy output, require us to think in terms of policy responses, not cures.

The emission of greenhouse gases is in quite a different category; it involves human decisions about investment and consumption. The first requirement in thinking about limits to these emissions is for real information about the problem. The relative energy-trapping propensities of the greenhouse gases is probably the only area of the topic which is relatively free of unknowns, since it depends upon repeatable laboratory measurements. The emissions of the main greenhouse gases are all more or less difficult to determine. Some, like CFCs and halons – chemicals related to CFCs, used in fire prevention – are produced and used in fairly precise amounts. Others are produced both by human activities and naturally, with the added complication that the natural rate of production can be altered by human activities, like the massive quantities of methane which may be produced in the Arctic if temperatures there rise.

The message which this sends to policymakers varies with the temperament of the hearer. However, the remarks made in this book about the long timescales involved and the need for more knowledge of the problem are not meant to be read as support for anyone who proposes standing in the way of the oncoming juggernaut on the grounds that it might turn out to be an optical illusion. Instead, a start must be made on the complex and difficult task of cutting greenhouse emissions. These come in two categories: carbon dioxide, and all the others.

The ragbag of "other" greenhouse gases contains at least one major group: the CFCs, which are already going severely out of industrial, political and consumer fashion after being implicated in the attack on the ozone layer. Their effect on the Earth's temperature will, as we have seen, take decades to work through. But a treaty has been drawn up to restrict their production – which, despite reservations about the agreement and the possible effects of the alternative chemicals which might be introduced once CFCs are phased out, implies that international action in this arena can work. Some authorities claim that all CFC use can be ended during the 1990s. It was also reported in 1990 that halons can be taken out of use within a few years, according to firms which use them on oil platforms in the British North Sea. Even if the sceptic credits the organizations involved with no sense of altruism, they seem in this case to be able to react to a potential problem before rather than after a public campaign to make them alter their behaviour.

Of the other major greenhouse gases one group, the oxides of nitrogen and nitrous oxide especially, are produced distinctively

Figure 7.2 **Contributions to global warming: (a) 1985; (b) 1980**

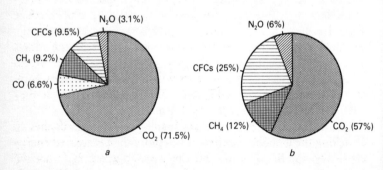

Source: Nature

by industrialized agriculture, mainly as a by-product of the use of nitrogen fertilizers. The use of such fertilizers is likely to be reduced progressively in the industrialized world in future years, and their sheer cost means that in the Third World they ought not be regarded as part of sustainable farm policies.

The cases of methane and carbon dioxide are more interesting and more intransigent. The example of methane from the Arctic shows that environmental damage from global warming can be self-feeding – stop pushing out the greenhouse gases and the potential production of more greenhouse gases will also be prevented. However, there is at the time of writing a debate about the extent to which it might be worth accepting more artificial emissions of methane to the atmosphere. The argument is that burning natural gas produces less carbon dioxide than coal or oil, so that the greenhouse effect is minimized, overall, by using more of it. Critics argue that gas pipelines leak so much that a worse green house effect is caused by using more gas than would result from burning something else instead. The answer may be to ensure that pipelines leak less, not to use coal instead of gas.

On carbon dioxide itself, the picture is no simpler. Too much of the world's economic activity is riding on fossil fuels for their use to stop. This implies that the priority is to implement energy policies, especially for developed nations, which reduce the use of fossil fuels, encourage their efficient use (in small cars, new-technology power stations and other efficient machines) and encourage substitutes like renewable energy. In a world with more powerful storms and winds, even wave and wind energy might start to make some of the progress which they have promised but not delivered for many years.

In the Third World, there are many possible approaches to attacking the greenhouse effect. Even more than increased energy efficiency, the first priority must be a real agreement to use the rich world's money to prevent attacks on tropical forests – or, at the very least, to end Western economic pressure for loan servicing which leads to their removal. This would eliminate a major source of new carbon dioxide emissions to the atmosphere, preserve a powerful machine for removing carbon dioxide, and bring a host of other human and environmental benefits.

It seems certain that the key to cutting carbon dioxide emissions in the end is simply to use less energy – choose transport options which use less power, especially public transport, make machines and buildings more efficient, and alter social structures to adjust to less energy-intensive ways of communicating, making things or enjoying ourselves. Conservation supporters claim that measures like this will allow us to overcome the greenhouse effect at a profit, since there are many energy efficiency measures which would more than pay for themselves in fuel savings. The same arguments also apply to less polluting forms of fossil fuel energy use, like novel coal-burning technology to increase power station efficiencies and reduce the amount of carbon dioxide emitted for each unit of electricity produced, or new vehicle engines to stretch fuel consumption. Even the supporters of nuclear power have tried to use the greenhouse effect in a campaign to prove that building more nuclear stations can cut carbon dioxide emissions, although the problems of economics, safety, public acceptability and environmental danger which have killed nuclear power in most countries will need more than the greenhouse effect to counteract them. By contrast, the arguments for energy conservation show that in some cases, action to slow global warming can reinforce strong arguments rather than being used to shore up failed ones. As well as saving money, energy conservation and energy efficiency have the power to slow global warming and its potentially horrific consequences from the poles to the equator.

All over the world, people governments, companies and organizations will be making policies and investments with consequences for global warming, and sea level rise, for decades to come. At every level, from international treaty-making to individual decisions about personal spending, such actions must take into account that ceasing to warm the Earth more than we have to will also slow the rising seas, and with it slow the possible loss of lands, countries, habitats and peoples without which we would all be poorer.

Canada

The government of Canada was one of the first in the world to take global warming seriously. Part of the work carried out by Environment Canada, a Federal agency, on the problem has been a pair of studies on just what will happen in two coastal areas of eastern Canada if predictions of a 1m rise in sea level turn out to be true. One study related to Charlottetown on Prince Edward Island; the other to a wider area, Saint John in New Brunswick and the lower reaches of the Saint John River.

This rise in. sea level would have effects including the complete inundation of some land, erosion of land in what are now inland areas, saline attack on rivers, land and groundwater, and increased flooding. The size of such effects depends upon everything from the detailed geology of aquifers to the profiles of river channels. Both the Charlottetown and the Saint John studies used sophisticated models to determine the possible effects, on the basis of detailed mapping and decades of tide data. The work concerned is probably a better guide to the full scale of the problem than many countries, especially in the Third World, have found it possible to produce.

The first study finds that in the City of Charlottetown itself, an array of expensive new waterfront developments would become uninhabitable, including a new convention centre and a courthouse. Several streets downtown would be below high-water mark or subject to flooding, and a total of 225 buildings would be vulnerable to floods. The sewage and storm drain systems would collapse during storms, and at high tide – and so would the swimming pool. Outside town, flood problems would affect road causeways and other structures. It may be that these flooding problems could be eased by building new protective works, and by altering planning practices to shift development away from vulnerable areas. Harder to solve would be the sea attack on sewage pipes and flood control systems.

The coastal area around Saint John could well be even more radically affected by sea level rise, since the town lies on the Bay of Fundy, which has some of the most spectacular tides in the world. The tidal range there is already 8.8m. A 1m rise in sea level would be accompanied by an increase in tidal ranges to perhaps 8.95m.

Higher tides and sea level would mean that two of Saint John's major residential suburbs would be inundated completely or subject to severe flooding. At the same time, the city's road and rail links would be seriously affected. There is a risk that links to the east of the city could be cut altogether. Further economic damage could be caused by the loss of key centres of employment, including the shipyard and sugar refinery. And as at Charlottetown, there is the severe risk that sewage and drainage systems built on the assumption of today's sea levels would be unable to cope. At Saint John there is a risk that the lagoons used to hold the city's sewage and industrial waste would be inundated. This would mean an immediate release of large amounts of pollution, and would also cripple the waste systems for some time. Other parts of Saint John which could be seriously affected by rising sea levels include the power station, which would be vulnerable to flooding, the city's fresh water supply, and the wharf area, which would be completely inundated.

These problems will probably be expensive to solve, and in the Canadians' view will call for changes throughout the planning process. However, potential problems within a single modest-sized city are at least geographically confined. The authors of the Canadian report note that there will also be severe effects along the whole of the Fundy coast. The effects on the Saint John river itself might be less sweeping because its size and shape are determined essentially by the fresh water coming down it, but even here, increased penetration of salt water might devastate local fish stocks. There would also be flood risks to prime agricultural land which would call for investment in protection and perhaps the adoption of new agricultural practices. Increased flooding could also cut the country's biggest arterial road, the Trans-Canada Highway, with severe effects on the whole road link between central and Atlantic Canada.

Bibliography

The literature on sea level rise is vast, and growing at a rate which threatens significant low-lying areas of the world with inundation. In the few weeks before completion of this book, publications bearing on the matter have ranged from technical books from learned societies concerned with geophysics to at least one football magazine (*The Absolute Game*, a Scottish publication which ran an excellent piece in its May 1990 issue on clubs due to be put out of business by the sea).

Sea level rise is touched upon in much of the literature on the greenhouse effect and global warming, which is too vast to be dealt with in detail here. Among useful general books are *Managing Planet Earth* (published by Freeman for Scientific American in 1990). The debate on climate change and the sea level rise it might cause is fought out mainly in international scientific journals like *Nature, Science* and *New Scientist*, where fresh views are aired on the matter each week. Too many articles to list from such learned sources have gone into this book.

In addition, awareness of the problem has been fed by a number of reports on its possible impact in particular regions and countries. To start with the developing world, examples include several studies on the South Pacific, like *Kiribati and Sea Level Rise* by R.F. McLean (Commonwealth Secretariat, 1989); *The Impact of Global Warming on South Pacific Islands Countries* (Ministry of Foreign Affairs, New Zealand, 1988); and *Potential Impacts of Greenhouse Gas Generated Climatic Change and Projected Sea-Level Rise on Pacific Islands States of the SPREP Region* (Association of South Pacific Environmental Institutions, 1988 – SPREP means South Pacific Regional Environment Programme); a 1989 report from the same group, *Studies and*

Reviews of Greenhouse Related Climatic Change Impacts on the Pacific Islands; and reports by James Lewis for the Commonwealth Secretariat, *Sea Level Rise: Tonga Tuvalu* (Commonwealth Secretariat 1988). The Secretariat also published Lewis's paper *Implications of Sea Level Rise for Island and Low-Lying Countries*, prepared in 1988 for its Expert Group on Climate and Sea Level Rise. A popular version of the SPREP work is *A Climate of Crisis* (Association of South Pacific Environmental Institutions, Port Moresby, 1989 – the precursor, I hope, of a wide range of forthcoming general publications on the subject). There has been a blizzard of analyses of the problem as it affects the Maldives, including *Maldives and Sea Level Rise* (Colin Woodroffe, University of Wollongong, Australia, 1989) and *The Implications of Sea Level Rise for the Republic of Maldives* (Alasdair Edwards, Commonwealth Secretariat, 1989); *Report of the Mission to the Republic of the Maldives* (UN Environment Programme, 1989); and *Republic of Maldives: Implications of Sea-Level Rise* (Ministry of Economic Affairs, the Netherlands/UN Development Programme, 1989). Work prepared for the Secretariat in support of the Expert Group includes *Effect of Climate Change and Sea Level Rise on Bangladesh* (1989) and *The Implications of Sea Level Rise for the Coastlands of Guyana* as part of the work of its expert group on climate change and sea level rise, along with Joanna Ellison of the University of California's look at *The Effect of Sea-Level Rise on Mangrove Swamps*. On the Mediterranean area (with sideways looks at the Caribbean and elsewhere) is a 1988 UNEP document, *Joint Meeting of the Task Team on Implications of Climatic Changes in the Mediterranean* (UNEP [OCA] WG 2/25).

On the USA there is *Greenhouse Effect, Sea Level Rise and Coastal Wetlands* (EPA, 1988); and the sea level rise chapters in *Preparing for Climate Change* (Government Institutes Inc., 1988). On Europe, Ted Hollis and his colleagues' paper *The Effects of Sea Level Rise on Sites of Conservation Value in Britain and North West Europe* was published by him at University College, London, in 1989. The journal *Ecos* was also due to carry a report on the same work. For Canada, *Preliminary Study of the Possible Impacts of a One Metre Rise in Sea*

Level at Charlottetown, Prince Edward Island (CCD 88–02) and *Effects of a One Metre Rise in Mean Sea-Level at Saint John, New Brunswick and the Lower Reaches of the Saint John River* (87–04) were published by the Canadian Climate Centre, Downsview, Ontario. On the Netherlands, "Will Climate Changes Flood the Netherlands?" by G.P. Hekstra was published in 1986 in *Ambio*, vol. 15, no. 6.

General books on sea level are rare, but the demand is met admirably by *Tides, Surges and Mean Sea-Level* by David Pugh (Wiley, 1987). Papers on the problem in general include the proceedings of the 1989 Small States Conference on Sea Level Rise, which included a wealth of country statements varying from the obvious to the detailed and insightful. Copies may still be available from the Ministry of Planning and Environment, Male' 20–05, Republic of Maldives; and the proceedings of the 1989 Cities on Water conference in Venice, available from the Cities on Water International Centre, S Marco 875, 30124 Venice, Italy. Other useful general documents include the documents for the 1989 10 Downing Street Seminar on Climate Change (Department of the Environment, London); chapter 5 of the 1990 *State of the World* yearbook, "Holding Back the Sea", by Jodi L. Jacobson (Norton); *The Impact of Climatic Change and Sea Level Rise on Ecosystems*, by the International Union for the Conservation of Nature and the World Conservation Union (IUCN, Gland, Switzerland, probably 1989); *Sea Level Rise, Consequences and Policies*, P. Vellinga (Delft Hydraulics, Emmeloord, Netherlands); *Greenhouse Warming and Changes in Sea Level*, Johannes Oerlemans (Institute of Meteorology and Oceanography, University of Utrecht, Netherlands); and *The Greenhouse Effect and Rising Sea Level*, R. A. Warrick and G. Farmer (Climatic Research Unit, University of East Anglia, Norwich, England).

Index

AT&T Bell Labs 22
Adriatic Sea 74
Allen, Hurricane 49
American Geophysical Union 29
Andaman Islands 42
Antarctic 18, 25–29, 112
Antarctic Convergence 114
Arabian Gulf 46
Arctic 14, 24, 25
asparagus 117
Assateague Island 41
Association of South Pacific Environment Institutions 96, 102, 105
Australia 42, 43, 56–57

Bakun, Andrew 111
Bangladesh 37, 42, 59, 68–70, 122–123
Banjul 123
Barbados 100–101
barley 117
Beijer Institute 21
Biuvakaloloma, Lt-Col Apolosi 102
Blackwater Wildlife Refuge 41
Brahmaputra, River 42, 68
Brooklyn Naval Yard 81
Brown, Barbara 46, 48

California 78
Canada 145–6
Cape Cod 78
carbon dioxide 6–12, 143–144
Caribbean Sea 43, 67–68
Caroline Islands 44
Charleston 78, 88
Charlottetown 145–146
Chlorofluorocarbons 6, 15–16, 141–142
Cities on Water 71, 123, 130
climate zones 50, 116
Common Agricultural Policy 85, 132
coral 32, 45–49, 56–57
coral bleaching 32, 47

CORINE 82, 83
Crown Agents 23

Dar-es-Salaam 123
deforestation 7–8
Delaware River Basin Commission 55
Delft Hydraulics 39, 59
Dhaka 68
Dominica 94
Dunwich 63

ERS–1 and ERS–2 108
Earth Radiation Budget Experiment 21, 110
Ebro, River 75
Edwards, Alasdair 61
Egypt 59, 65
Elbe, River 72
Elena, Hurricane 38
Ellison, Joanna 43, 44, 45
energy conservation 144
Environmental Protection Agency 55, 78, 87, 103
Everglades, Florida 42, 122
European Commission 82
European Regional Development Fund 136
European Space Agency 113
Exclusive Economic Zones 95–96

Fiji 99–100, 105
Filchner Ice Shelf 25, 26
flood protection 120–123
France 76
Frassetto, Roberto 123
Fundy, Bay of 145–146

Gambia 59, 123
Ganges, River 42, 68
Georgia (USA) 41
Gilbert Islands 94
glaciers 26, 29

Global Sea Level Observing System 113, 114
Grand Cayman 43, 44
Great Barrier Reef 56–57
Greenhouse Effect 2, 5–17, 138–144
Greenland 25, 26, 112
Gulfport, Mississippi 38
Guyana 50, 124–125

halons 16, 141
Hamburg 38, 72, 119
Hawaii 8, 9
Hekstra, G P 89
Himalayas 68, 70, 123
Holderness 87
Holdgate, Martin 60
Hollis, Ted 82
Holmes, Patrick 61
Hopa, Jack 103

Ibe, A Chidi 123
Ichkeul National Park 67
India 42
Indonesia 38, 59, 65
International Programme on Climate Change 122
International Union for the Conservation of Nature and Natural Resources (IUCN) 49, 60
Ireland 37
Isaac, Hurricane 94
Italy 65

Jaeger, Jill 21
Jaywick Sands 121

Kalar 117
Kiribati 98, 99, 105
Kiritimati 98
Krakatoa 38
Kuo, Cynthia 22

Lagos 123
Leatherman, Stephen 40–41
Leningrad 73
Lewis, James 93, 94, 95
limestone 10
Lindzen, Richard 109–110
Little Ice Age 18, 23
London 3, 18, 73

Louisiana 40, 55, 86
Lyon, Gulf of 76

Maine 78
Majuro 96
Maldives 4, 32–35, 59, 92, 96, 103
Male' 32, 34, 92, 96
mangroves 42–45, 118
Marshall Institute 24
Marshall Islands 96
marshlands see wetlands
Mataku 94
Mauna Loa 8, 9
McLean, R F 98
Mediterranean Sea 65–67, 73–77
Meier, Mark 29, 30
Melanesia 97
Mersey, River 73
methane 6, 12–14, 143
Meteorological Office 114
Mexico 46
Mikolajewicz, Uwe 114
Mississippi, Delta 78
Mozambique 59
Myers, Lincoln 101

National Aeronautics and Space Administration (NASA) 22, 113
National Oceanic and Atmospheric Administration (NOAA) 111, 114
National Research Council 117
National Trust 82
Natural Environment Research Council 37, 114
Nauru 96
Netherlands 3, 37, 85, 88, 90–91
New Caledonia 96
New Guinea 96
New York 81
New Zealand 96, 106
Newfoundland 145
Nigeria 123
Nile, River 65
Nino, El 48
nitrogen oxides 6, 14–15, 142–143
North Carolina 38, 80
Nufu'alofa 105
Nurse, Leonard 100

Office of Technology Assessment
    (OTA) 4
Oerlemans, Johannes 25, 26
Osaka 123
ozone 6, 16, 17

Pacific Ocean 104–106
Pakistan 59
Papua New Guinea 42, 95, 105, 106
Pendine Sands 121
Pernatta, John 95, 96, 97
Philadelphia 55
polar "platform" 108
Prince Edward Island 145
Po, River 65
ports 62–63, 79–81
Pugh, David 37, 38, 42, 74

Quandong 117

Radick, Richard 24
Ramanthan, A 21
Rawal, A 21
Rhine, River 91
rice 117
Ross Ice Shelf 25, 26
Ross Sea 26, 115
Rotterdam 91

St Helena 37
St John 145–146
Salicornia 117
salt damage 51–56
salt resistance 117–118, 132–133
Senegal 59
Sites of Special Scientific Interest
    (SSSI) 82, 85
Solomon Islands 106
Somerset Levels 40
South Carolina 41, 88
Spain 75, 83–84
storm surges 38–39
sugar cane 122
Sumatra 42
Sun 23–24
sunspots 23
Surinam 59

Tambora 110
Tans, Pieter 22

Tanzania 123
taro 34, 54
Temperature (of the Earth) 18–22
Texas, Gulf of 39–40
Thailand 59
Thames Barrier 3, 121, 129
Thames, River 73
Thatcher, Margaret 8
Thermaikos, Gulf of 76
thermal expansion 24, 30
Thessaloniki 76–77
tides 36–37
Titus, James 55, 78, 103
Tokelau 96, 105
tomatoes 117–118
Tonga 43, 44, 94, 103, 105, 106
Tongatapu 44
Towyn 120
Trinidad and Tobago 101
tsunami 38, 94
tundra 14
Tunisia 67
Tuvalu 105

UN Environment Programme 10, 12,
    13, 15, 16, 17, 30, 67, 74, 75
UK 82–87
USA 77–81, 87–89
US National Meteorological Centre
    21
Uiha 94

Vanuatu 103
Vellinga, Pier 59
Venice 63, 73–75, 123, 129
Venus 6, 10
Virginia 40, 41
volcanoes 14

Warrick, Richard 112
water supplies 51–52
water vapour 20, 111–112
Wedell Sea 26
Wegener Polar Studies Institute 25
wetlands 41, 51, 78–79, 88–89
wheat 117
Wigley, Tom 20, 21, 24
Woodroffe, Colin 52

Yorkshire 87